ESSENTIAL PHYSICS, CHEMISTRY AND BIOLOGY

A guide to important principles
for nurses and allied professions

Essential Physics, Chemistry and Biology

D. F. Horrobin

Published by
MTP Press Limited
Falcon House,
Cable Street,
Lancaster.

ISBN 0 85200011 1

First published 1971

Reprinted 1975, 1978

Printed and Bound in Great Britain by
A. Wheaton & Co. Ltd., Exeter

Contents

1

Physics, chemistry, and biology. Introduction

To a considerable extent our modern understanding of medical science is based upon knowledge of physics, chemistry and biology. A sound appreciation of certain aspects of these subjects can make the learning of more obviously medical topics much easier and can lead to a deeper understanding of modern medicine.

Yet these three subjects, and in particular physics and chemistry are often thoroughly unpopular with both nurses and medical students. The reasons are not hard to find. School science courses are naturally not orientated towards medicine and are often taught by people with little knowledge of medicine. Aspects of interest and relevance for future nurses and doctors are not emphasized and some topics which are important are not even discussed. There is nothing reprehensible in this for schools could not orientate their courses to narrow professional ends. Nevertheless it does unfortunately mean that most nurses and medical students regard physics, chemistry and biology as dry-as-dust subjects which must be quickly learned and as quickly forgotten before going on to the real business of nursing or medicine.

This book attempts to show that some aspects of these subjects can be interesting, intelligible and very relevant to both the theoretical and practical aspects of learning to be a nurse. It does not attempt to cover any formal course in physics, chemistry or biology. Most aspects of the three subjects are ignored completely and the emphasis is on those topics of direct concern to nurses. The aim of the book is to enable nurses to achieve greater understanding of such clearly 'medical' subjects as biochemistry, physiology medicine, drug action and anaesthesia. It should be read before attempting any of these subjects and should be kept constantly at hand so that it can readily be consulted when difficult topics arise. I hope that it may succeed in reducing the fear with which many nurses face the sciences with which the book deals.

Section 1

BIOLOGY

2

The cell and its requirements

The world of living things is conveniently and conventionally divided into two great groups, the animals and the plants. Broadly speaking the important feature which distinguishes plants is that they can manufacture most of the substances they require by trapping and using various forms of outside energy, in particular the energy of sunlight. In the process of photosynthesis they utilize the energy of light to build up complex chemical substances from relatively simple ones.

In contrast, animals lack the ability to use light or any other form of outside energy. Instead they must obtain the energy they require by breaking down complex substances which ultimately they always obtain from plants. Plant-eating animals such as cows and sheep obtain these substances directly. Carnivores obtain them indirectly after they have passed through the bodies of other animals.

During the past few decades, the properties of a third group of organisms have been determined. These are the viruses and they are particularly important in the causation of many human diseases such as smallpox, polio and measles. They are unusual in that while they can survive outside the living cells of animals or plants they can only reproduce themselves while they are parasites actually inside plant or animal cells. They use the mechanisms which they find within the cells for the manufacture of more virus material.

The concept of the cell

All plants and animals are made up of very large numbers of subunits known as cells. In both groups of living organisms the simplest species consist of single cells. We can learn a great deal about the functioning of such complex creatures as human beings by studying the structure and the requirements of a relatively simple single-celled animal such as the familiar amoeba.

All animal cells have certain structural features in common. The

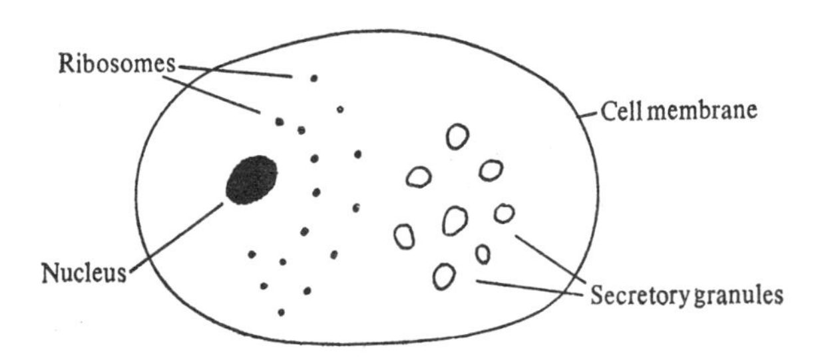

Fig.2.1. An outline sketch of a cell.

three most important of these are the cell membrane, the nucleus and the cytoplasm. The cytoplasm is the material within the cell membrane but outside the nucleus. The cell membrane is an extremely important structure about which as yet we know relatively little. It is not an inert barrier which lets virtually anything in or out of the cell. It is highly selective and exerts a very active control over substances which try to cross it. Some substances, like sodium, for example, are kept out of the cell and if they try to gain entry they are forcibly ejected. Potassium on the other hand is kept inside the cells. As a result the concentration of potassium inside cells is high and that outside cells is low. The reverse is true of sodium. We are only just beginning to understand the ways in which the cell membrane functions and full knowledge lies far in the future.

The other universally found structure is the nucleus. This contains the chromosomes which seem to consist mainly of deoxyribonucleic acid or DNA (see books on biochemistry and medical genetics). The nucleus is essential for the normal development of the cell because the chromosomes contain the plans for the manufacture of all the substances and structures required within the cell. It is also involved in the day-to-day running of the cell's activities. A breakdown of this control exerted by the nucleus can probably account for many diseases which as yet are poorly understood, and in particular, cancer.

The cytoplasm

This is the part of the cell outside the cell nucleus but within the cell membrane. It varies considerably in its structure depending on the function of the cell. In most cells the cytoplasm contains a series of folded membranes on which are tiny granules known as the ribosomes. In the ribosomes the manufacture of the proteins which the cell needs takes place. In cells which secrete substances, such as cells which manufacture the digestive juices or the hormones, there are often other granules in the cytoplasm known as secretory granules. These contain the secretions wrapped up in what appear to be little packets. At the appropriate time the cell can push them out across the cell membrane, so releasing their contents. In most cells which secrete things there is a complex structure, the Golgi apparatus, which is thought to be the place where the secretory granules are manufactured.

Cells with a 'brush border'

The brush border is a structure which is found in many of the cells which line the gut. It is found only on one side of each cell, the side which faces the gut cavity. When looked at with an ordinary microscope it looks as though the cell has a brush-like structure on that side, hence the name 'brush border'. But when looked at under the

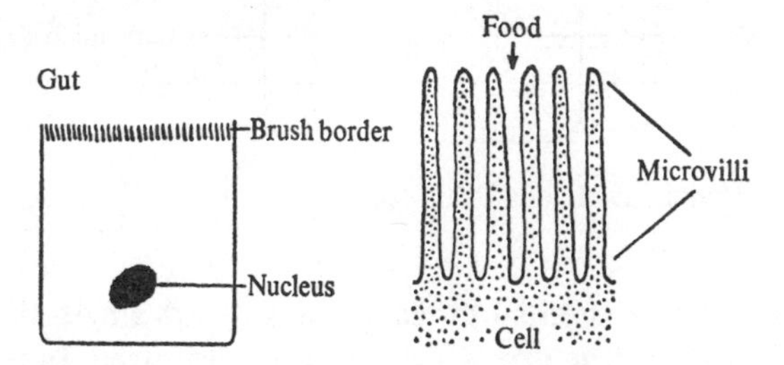

Fig.2.2. Cell from the lining of the gut showing the brush border. The right-hand diagram shows the border greatly magnified.

much greater magnification of the electron microscope, it can be seen to consist of innumerable finger-like processes known as microvilli. The function of the microvilli is greatly to increase the surface area of the cell. You can visualize the effect by comparing the surface area of a desk top with the surface area of the same desk when a lot of pencils are standing on it upright and close together. The surface area of the desk top plus the surface area of all the pencils is much greater than the surface area of the desk alone.

The microvilli thus greatly increase the surface area of the cell membranes facing the gut cavity. The reason for this is that all food must be absorbed from the gut by crossing the barrier of the surfaces of the lining cells. Each fragment of surface can only absorb food at a limited rate and so the rate of absorption can be greatly increased by increasing the amount of surface area available. In some diseases such as sprue and coeliac disease the intestinal wall is damaged and many of the cells lose their microvilli. As a result food cannot be absorbed properly and passes straight through the gut into the faeces causing diarrhoea. Because relatively little of the food can be absorbed, the patient effectively becomes starved. The general term for this condition is the malabsorption syndrome.

The requirements of a single-celled animal

All single-celled animals live in water, in ponds, in rivers, in lakes

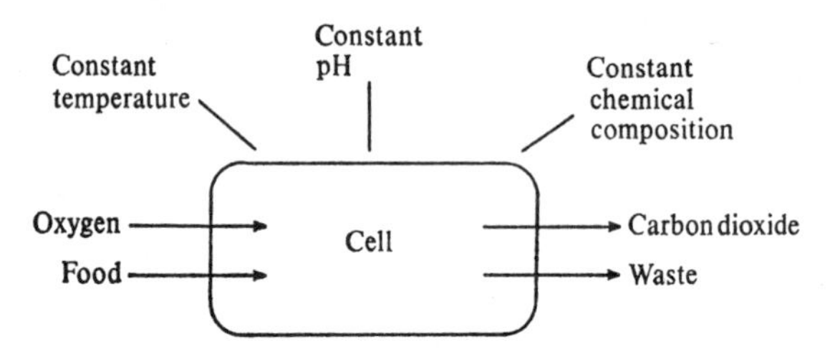

Fig.2.3. The requirements of a living cell.

or in the sea. It is essential to recognize that such single animal cells are very delicate structures which are easily destroyed. In particular, they can be damaged by three types of changes in the fluid which bathes them.

1. Changes in temperature. They can be killed if the temperature of the water becomes either too high or too low.

2. Changes in the acidity or alkalinity of the fluid. Most of the water in which such animals live is nearly neutral, being neither very acid

nor very alkaline. If by chance the water does become acid or alkaline, the animals within it are often killed.

3. Changes in the chemical composition. Each animal tends to be adapted to a particular chemical composition. For example, animals which live in the sea are accustomed to a high salt content and are killed if they are placed in fresh water. Conversely most freshwater animals are killed if they are placed in the sea. Each animal is therefore adapted to a particular chemical composition of the surrounding fluid and can be killed if this changes very much.

Most single-celled animals do not have to worry about these problems. The temperature changes of the water in which they live are much smaller than the temperature changes on land. The chemical composition and acidity of most lakes and rivers, and of the sea is virtually constant. Single-celled animals do not therefore have to make any attempt to regulate the environment in which they live.

Apart from requiring constancy in certain environmental conditions, animal cells need to obtain certain things from their surroundings. The most important of these are food materials and oxygen. Food is obtained by engulfing tiny particles of animal or vegetable origin floating in the water. The process is known as phagocytosis and the food is taken into the cell across the cell membrane. Some of the cells in the human body are capable of phagocytosis. The most important are some of the white cells in the blood and some of the cells which line blood vessels in the liver, the spleen and the bone marrow. Their function is to remove unwanted particles of solid matter from the blood, ranging from bacteria to old and dying red blood corpuscles.

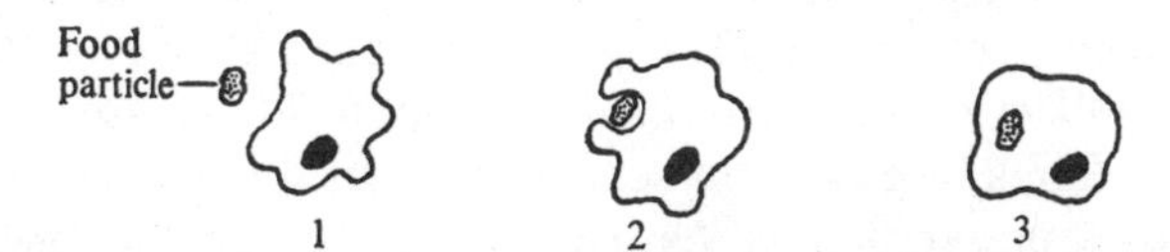

Fig.2.4. The process of phagocytosis by which some types of cell engulf solid particles.

Oxygen is essential for the processes in which food is broken down to supply energy. It also is taken in directly from the surrounding water, this time by the mechanism known as diffusion. Diffusion is very important in all living creatures, including humans, and it is important to understand it. Diffusion occurs because all substances tend to move from regions where they are in high concentration to regions where they are in low concentration: the process continues until no concentration differences remain. This may very simply be

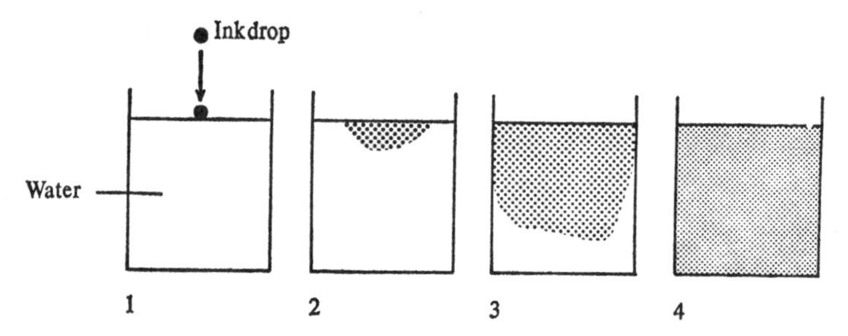

Fig.2.5. When an ink drop is placed in a beaker of water the ink soon spreads evenly through the water because of diffusion.

illustrated by putting a drop of blue ink into a jar full of clear water. At the instant of putting the drop in, the concentration of the blue dye is very high in the drop and zero in the water. But as soon as the drop enters the water, even if the water is absolutely still and is not stirred, the dye begins to spread through the water from areas where there is a lot of dye (where it is high in concentration) to areas where there is little or no dye (where it is low in concentration). Eventually by the process of diffusion the dye becomes evenly distributed throughout the water and there are no concentration differences left.

The law of diffusion does not apply only to ink and water. It applies to all gases and to all liquids. All substances in gases and liquids tend to move from areas where they are high in concentration to areas where they are low in concentration. In the case of a single animal cell in a lake or in the sea, the cell uses up oxygen and therefore the concentration of oxygen within the cell falls below its concentration in the surrounding water. There is therefore a concentration difference between the oxygen level in the cell and the oxygen level in the water. Oxygen therefore diffuses from the water into the cell. Precisely the same thing happens in the human body. Cells use up oxygen and the oxygen concentration within them falls below the oxygen concentration in the blood. As a result oxygen moves from the blood into the cells.

Excretion of substances

Single cells do not only take things from the water in which they live. They also pass things out into the surrounding water. The burning of food to give energy yields carbon dioxide and many other waste products such as sulphuric acid and ammonia. As these substances are produced, their concentration inside the cell rises above their concentration in the surrounding water. They therefore move out of the cells into the water by the process of diffusion. They do not usually significantly alter the composition of the water because the volume of that water is so vast compared to the volume of a single cell. The waste products therefore rapidly become diluted. Occasionally this is not true. For example, under certain conditions tiny animals known as dinoflagellates can multiply in the sea in vast numbers. They are reddish in colour and there may be so many of them that the sea as a whole appears to be red. They produce some highly poisonous waste products and these can destroy all other living things in the vicinity and make animals like shellfish too toxic for human beings to eat. However, this sort of thing is very unusual and single-celled organisms normally cannot alter to any significant extent the composition of the water in which they live.

The need for movement

Plants can usually obtain all they need without moving. They utilize the energy of the sun and collect the very simple chemicals

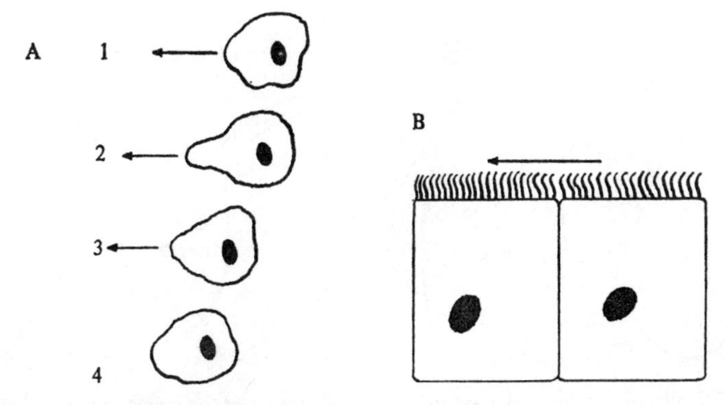

Fig.2.6. A. Amoeboid motion. B. The use of cilia to generate a current of fluid passing over a cell surface.

they require from the surrounding water or from the soil. But animals have to move in search of food. Tiny single-celled animals go about this in two different ways. Some like the amoeba move by pushing out part of their cell membrane and then allow the contents of the cell to flow into the part which has been pushed out. This is known as amoeboid movement and it is also important in the human body. Some of the white cells of the blood which are essential for the protection of the body against bacterial invasion move through the tissues in an amoeboid fashion and then engulf the bacteria by means of phagocytosis.

Some other single-celled animals are covered by masses of tiny hair-like structures known as cilia. These thrash the water in a co-ordinated way generating currents which push the water backwards and the animal forwards. Again, cilia are important in humans but not in quite the same way. Many human cells which line the internal surfaces of the body such as the respiratory tract or the Fallopian tube which goes from the ovary to the uterus have cilia on their surfaces. The cells are fixed in position and lashing of the cilia cannot move them. But the ciliary motion can move fluid lying on the surface of the cell. Thus the cilia in the respiratory tract generate currents which steadily wash mucus, dust, bacteria and many other types of material out from the lungs into the throat where they may be either coughed up as sputum or swallowed and destroyed by the secretions of the digestive tract. This process is vital for keeping the lungs clean. Some poisonous gases paralyse the cilia. As a result all the material cannot be removed from the lungs and the patient

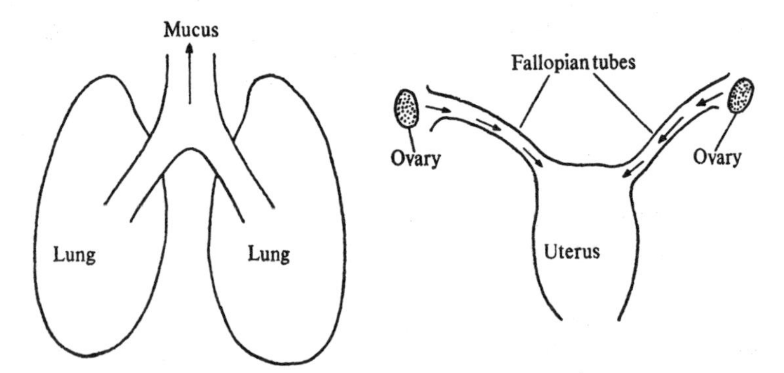

Fig.2.7. Places in which cilia are important. In the lungs they remove mucus, dust and bacteria. In the Fallopian tubes they generate a current which washes the ova down to the uterus.

often dies quickly from pneumonia. In the Fallopian tubes, the cilia generate currents which steadily wash the secretions of the cells lining the tubes down from the ovaries into the uterus. This current of fluid takes with it the eggs and ensures that they arrive safely in the uterus.

Reproduction

If living species are to survive, the individual members of that species must reproduce themselves. Single-celled organisms usually do this by a simple process of cell division but this is obviously impossible for larger organisms. Most larger organisms, including of course humans, reproduce themselves by some form of sexual activity. The important features and consequences of this type of reproduction are discussed in the book on medical genetics.

3

The functioning of larger animals

It is obvious that large animals consisting of millions of cells face many more problems than single-celled organisms. Even in large animals, however, each single cell is bathed in fluid, usually known as the extracellular or interstitial fluid. This fluid is not in

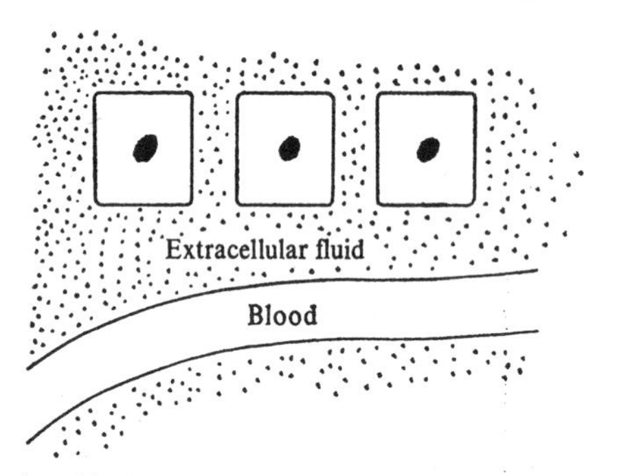

Fig.3.1. The relationship between the blood, cells and extracellular (interstitial) fluid.

free and easy communication with the surroundings of the animal. This chapter is concerned with the ways in which large animals, including human beings, cope with the difficulties posed by their size.

Large animals which live in water

Living on land introduces even further complications, and so first of all we shall consider a large animal which lives in water such as a fish. Perhaps the most obvious feature of a large creature like this which distinguishes it from a single-celled animal is that there are millions upon millions of cells in the body and each cell does not carry out all the functions which the body as a whole needs. In the amoeba, the same single cell collects food, oxidizes the food to supply energy, moves about, excretes waste products and is responsible for reproduction. In larger animals, each cell is specialized to carry out only some small aspect of the functioning of the body as a whole. Cells which are specialized to carry out a particular function tend to be gathered together in groups forming the various organs of the body.

The whole cell surface of the amoeba is in contact with the surrounding water. The cell can easily take from its surroundings what it needs and discharge its waste products directly into the lake or into the sea. In a fish, most of the cells lie deep within the body, far from the water in which the fish swims. Most of the cells in a fish cannot directly obtain food and oxygen from that water, nor can they discharge carbon dioxide, ammonia and other waste material into it. The cells in a fish, it is true, are surrounded by the water of the extracellular fluid: it is from this fluid that they must obtain their oxygen and food and into this fluid must pass their carbon dioxide and other waste.

The circulatory system

In both the fish and the amoeba, each cell is surrounded by water. The main difference is that the volume of water surrounding an amoeba in a lake, a river or the sea is vast in comparison to the size of the amoeba. It is therefore impossible, except in very rare circumstances, for single-celled organisms to alter the composition of the water which surrounds them by what they take out of it or put into it.

The situation in large animals is very different. The volume of the extracellular fluid surrounding each cell is very small. In most cases the volume of this fluid is much smaller than the volume of fluid actually within the cell. A living cell can therefore very rapidly exhaust all the oxygen and food in the fluid immediately surrounding it. Such a cell can also rapidly alter the acidity and chemical composition of the surrounding fluid. Most cells produce quite a lot of heat and the temperature of the fluid can therefore rise. In fact any living cell, if it had to make do solely with the extracellular fluid immediately surrounding it, would very quickly kill itself by using up the oxygen and poisoning itself with waste products.

This is why it is so essential to have a circulatory system which can bring oxygen and food to the extracellular fluid and can take away

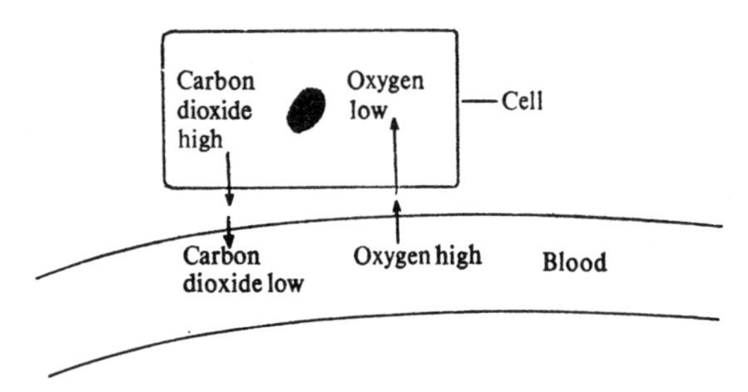

Fig.3.2. Cells generate carbon dioxide and use up oxygen. Carbon dioxide therefore leaves the cells for the blood while oxygen leaves the blood for the cells.

carbon dioxide, heat and other waste products. The way in which the circulation works can be illustrated with reference to oxygen and carbon dioxide. A living cell continually uses oxygen and this lowers the concentration of oxygen inside it below that of the extracellular fluid. Oxygen therefore flows by diffusion from the extracellular fluid into the cell. Because the volume of the extracellular fluid is so small it, too, is rapidly depleted of oxygen and its oxygen concentration falls well below the oxygen concentration of the blood. Oxygen therefore diffuses from the blood into the extracellular fluid. The cells are supplied with oxygen because of the process of diffusion. Similarly, living cells continually produce carbon dioxide. This raises the carbon dioxide concentration of the cell above that of the extracellular fluid and carbon dioxide therefore diffuses out of the cell into the fluid. In turn the carbon dioxide concentration in the extracellular fluid rises above that in the blood and so the carbon dioxide passes into the blood which carries it away.

It is clear that if the circulatory system is to work effectively it must bring the blood into close contact with the extracellular fluid surrounding every single cell in the body. It does this by an extremely complex system of blood vessels, the smallest of which are known as capillaries and have walls only one cell thick. There are capillaries in the immediate vicinity of every cell and it is across their walls that the exchange between the blood and extracellular fluid takes place. The blood is of course continually pumped around the circulatory system by means of the heart.

The constancy of the internal environment

If cells in a large animal are to survive, it is essential that the temperature and composition of the fluid immediately surrounding them must be kept constant. In the words of a famous French physiologist, Claud Bernard, the constancy of the internal environment of the cells is the main problem facing animal life. As we saw in the previous section, this constancy is maintained by the use of the blood which brings oxygen and food to the extracellular fluid and takes carbon dioxide, heat and other waste substances away.

But the blood itself has only about one eighth of the volume of all the cells in the body and so, like the fluid surrounding the cells, it too is in danger of being rapidly depleted of its oxygen and food materials and of having its acidity, temperature and chemical composition drastically altered by the waste products of the cells. In fact, of course, the temperature and chemical composition of the blood remain remarkably constant. This is because the blood itself is served by a number of organs whose function is to maintain the constancy of its temperature and composition. The important organs which do this in a creature like a fish which lives in water are the following.

1. THE ALIMENTARY TRACT OR GUT. Food is taken in at the mouth and passed along the alimentary canal where it is first broken down into tiny particles (digested) and then taken into the blood across the gut wall (absorbed). This ensures that the blood receives a supply of food material.

2. THE LIVER. The supply of food is usually intermittent, and if the blood relied entirely on the food absorbed from the gut there would be long periods when no food was being taken in when the concentration of food materials in the blood would be in danger of becoming very low. This situation is partially counteracted by the liver and some other organs which store food when it is plentiful and is being rapidly absorbed from the gut and which then release some of their stores when the gut is empty and no food is entering the blood that way.

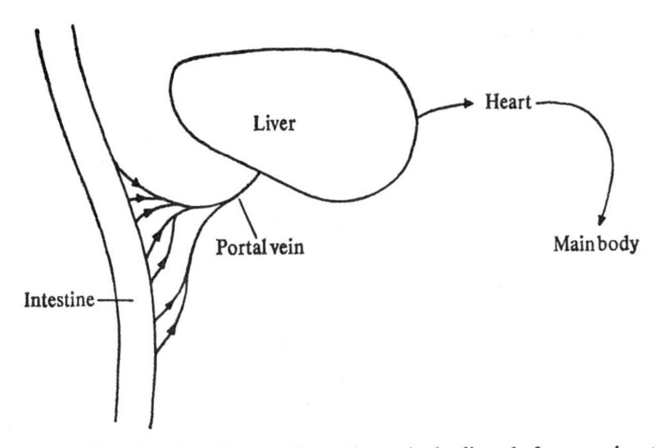

Fig.3.3. The blood leaving the gut flows through the liver before passing to the heart and out to the rest of the body. During meals the liver removes excess food from the blood and stores it, releasing it again when required.

3. THE RESPIRATORY SYSTEM. In fish this consists of the gills, and in land animals of the lungs. In the gills the blood comes into close contact with a continual stream of water drawn from the surrounding environment and which has a relatively high level of oxygen and a low level of carbon dioxide. Oxygen therefore diffuses in and carbon dioxide diffuses out, so keeping the oxygen and carbon dioxide levels of the blood constant. The lungs are continually supplied with large amounts of fresh air in the process of breathing. Air has a high oxygen content and a very low carbon dioxide content, and so land animals too can obtain oxygen from and lose carbon dioxide to their surrounding environment.

4. THE KIDNEYS. These paired organs are responsible for the excretion of acid, ammonia and other waste materials. In fish these substances can be simply poured into the surrounding water but in land animals they must be stored in the bladder which can then be emptied at intervals.

Special problems of man and other land animals

Animals which live in water, even those which are very large, have three important advantages over those which, like man, live on land.

The land animals must find ways of solving these three major problems.

1. Water-dwelling animals have an unlimited supply of water, whereas land animals obviously do not. A previously healthy man will die within 2 or 3 days if he is totally deprived of water. If he has water, but is totally deprived of food, he can survive for several weeks. Water is therefore of much more immediate importance than food, and land animals must always ensure that they have adequate supplies for drinking.

2. Water-dwelling animals can excrete their toxic waste materials directly and immediately into the surrounding water, which then dilutes the waste matter and renders it harmless. Land animals cannot do this. They must excrete their waste in a relatively small volume of urine and they must store their urine for periods before it is excreted. They cannot therefore afford to excrete highly toxic materials directly. Instead land animals must quickly detoxify such materials and thus render them harmless. This particularly applies to ammonia, a highly dangerous product of protein metabolism. In land animals this must rapidly be converted by the liver to the relatively harmless substance urea. Only then can it safely be excreted in the urine.

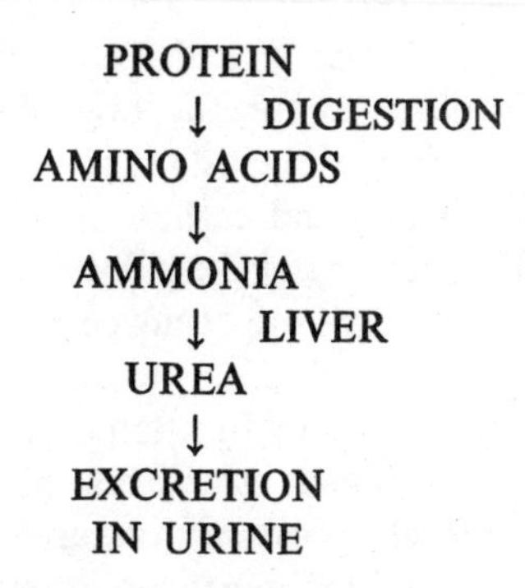

Fig.3.4. The formation of urea from amino acids.

3. Temperature regulation. The rates of the chemical reactions on which the working of the whole body depends are very much influenced by temperature. When the temperature falls, the reactions become slower and slower and an animal becomes more and more

sluggish. When the temperature rises, the reactions go faster and faster and if they become too rapid the animal may die. Animals which live in water and particularly those which live in the sea are largely protected from violent temperature changes since the temperature of water never rises so high or falls so low as that of the land.

Land animals face much bigger problems. Those whose body temperature depends on the temperature of the environment, the so-called poikilotherms or cold-blooded animals, cannot function at all in winter because their body temperature never rises high enough. They therefore remain inert for half the year. Two groups of animals, the mammals to which man belongs and the birds, have solved this problem by making their body temperature independent of that of the environment. They are the so-called homoiotherms or warm-blooded animals which maintain a constant blood temperature, in the case of man in the region of 37 °C or 98 °F. This constancy can be achieved only by having complex mechanisms which warm the body when it is too cold and which cool it when it is too hot. These mechanisms are discussed in the physiology book. They enable the warm-blooded animals to function effectively all the year round.

The maintenance of a constant internal environment

All the systems of the body, if they are to function effectively, must be subjected to some form of control. Food intake must be controlled by employing the device of appetite, otherwise too much or too little food would be eaten. The breathing rate must be controlled so that oxygen can be supplied and carbon dioxide removed at precisely the right rate. The working of the kidney must be controlled so that water and waste materials are removed neither too slowly nor too quickly.

The precise control of body function is brought about by means of the operation of the nervous system and of the hormonal or endocrine system. The most important structures in the control system as a whole are the brain and the spinal cord, often together known as the central nervous system or CNS.

The most important thing to note about any control system is that before it can control anything it must be supplied with information. Before the CNS can decide whether or not a person should be hungry, it must be supplied with information about the level of food materials present in the blood. Before it can decide whether a person should breathe more slowly or more quickly it must know how much oxygen and carbon dioxide is present in the blood. The same is true for the control of the kidney and of the activity of all other organs in

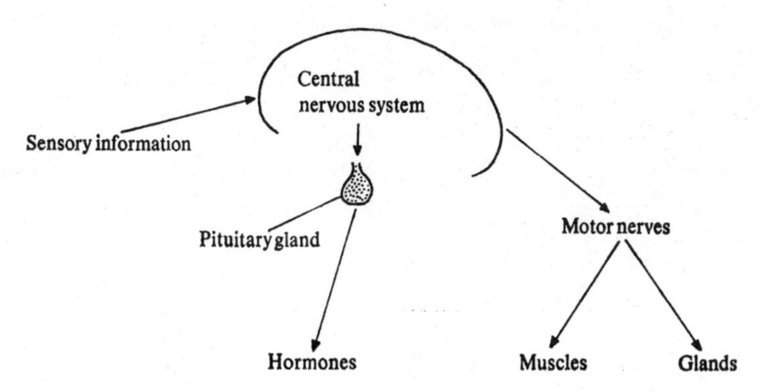

Fig.3.5. Outline of the body's control systems (see text).

the body. Therefore the first essential in any control system is an adequate system of collecting information about the state of the body. This is done by means of a complex system of sensory receptors whose function is to sense the state of all aspects of bodily function. There are receptors for blood glucose, for oxygen, for carbon dioxide, for the body water content, for blood acidity, for body temperature and for a myriad other things. Some of these receptors lie within the brain itself. Some are at the ends of long nerve fibres known as sensory nerves which transmit the information they collect back to the CNS in the form of nerve impulses. In this way the CNS can build up a comprehensive picture of what is happening all over the body.

Once the CNS knows what is happening, it must then have a means for rectifying the situation if something is going wrong. There are two available methods for doing this, by using nerve fibres and by using hormones. The motor nerve fibres are long, thin fibres which carry instructions from the CNS to the muscles and glands throughout the body. For example, if breathing is too slow so that the oxygen content of the blood is falling and the carbon dioxide content is rising, then the motor nerves can tell the muscles of the diaphragm and chest wall to move more rapidly and more deeply so returning the situation to normal.

The other way that the CNS can act is via hormones. Hormones are chemicals which are released by glands known as endocrine glands. The hormones are secreted into the blood and are carried by the circulatory system all over the body. As far as hormones are

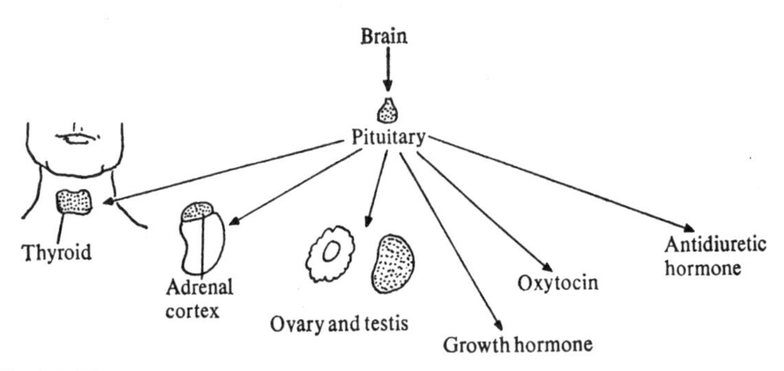

Fig.3.6. The actions of the pituitary gland.

concerned the brain acts via the pituitary gland, a minute structure lying in an inaccessible position in the centre of the skull at the base of the brain. The pituitary secretes a large number of hormones which are carried by the blood to every part of the body. The rate of secretion of each one of these is under the direct control of the brain. Some of the pituitary hormones act directly in their own right: growth hormone which helps to control growth and the level of blood glucose, and antidiuretic hormone (ADH) which governs the amount of water put out by the kidneys are examples of this type. Other pituitary hormones, however, are known as trophic hormones because they exert their effects by altering the behaviour of other endocrine glands throughout the body. For example, thyroid-stimulating hormone (TSH) from the pituitary, controls the behaviour of the thyroid gland which in turn governs the rate of working of many chemical reactions. Adrenal corticotrophic hormone (ACTH) from the pituitary, controls the behaviour of the adrenal cortex which in turn itself controls many different aspects of the function of the body.

By using the nervous and endocrine systems, the body is therefore able to control the behaviour of all the organs which help to maintain the internal environment of the body constant. Only if this constancy is maintained can the cells survive. If the regulating mechanisms fail, the fluids become poisoned and the cells and the organism as a whole die.

Support and movement of large animals

Large animals, like man, must have a system of supporting the body so that it does not collapse under its own weight. This is achieved by the possession of a bony skeleton to which all the soft tissues of the body are ultimately attached. But it is also essential to be able to move the body easily and so it is no use having an absolutely rigid central support. Instead the skeleton must have many individual bones connected together by flexible joints. In order to move the joints, there must be energy generating structures—the muscles—which pull on the bones and thus move the body in appropriate ways. It is possible to ensure that movements are appropriate only by controlling them effectively. This is the job of the CNS. The sensory nerves constantly supply it with a barrage of information about the position of the body and the state of contraction of the muscles. The CNS then sends instructions out along the motor nerves. In this way the movements of the body are made appropriate to its needs.

Section 2

CHEMISTRY

4

The structure of chemical substances

Chemistry is a subject rarely viewed with favour either by nurses or by medical students. Most regard it as a chore, irrelevant to medicine, which must be got over before the really interesting part of studying can begin. It is undoubtedly true that the chemistry can be taught in a dull and perhaps irrelevant way. This will happen, in particular, if the teacher does not really know about the medical applications of chemistry. As a result, subjects which do matter may be largely ignored while ones which do not matter at all are gone into in great depth.

This is a great pity because it is increasingly clear that many diseases can be understood in terms of disorders of bodily chemistry. Diseases of the endocrine glands such as diabetes or thyrotoxicosis, diseases of the bones and diseases of the kidneys cannot be properly appreciated without some knowledge of chemistry. Nor without chemistry can the administration of intravenous fluids, now so important in hospital, be understood. But fortunately the depth of chemical knowledge required is not very great. The account in this book attempts to provide that amount of chemical knowledge which is essential for a basic understanding of disease. It may disconcert some chemistry teachers who happen to read it, but for nurses I hope that it will illumine many areas which would otherwise remain dark.

The elements

First of all it is important to have a rough idea of the fundamental structure of the materials of which our bodies are made. The earth consists of about one-hundred primary substances, or elements.

An element is a substance which chemically cannot be further split up to yield simpler substances: it is a basic building-block of the material of the earth. Elements are sometimes found in the pure state but more often they are mixed or combined with one another in various ways. Gold, silver, carbon, oxygen, iron and calcium are all examples of elements and none of these can, by ordinary chemical methods, be split up to yield further simpler substances. Recently in nuclear reactors it has been demonstrated that one element can be converted to another, but that is a development which need not concern us here.

Fortunately, of all these elements, only a very few are important when studying the human body. Of overwhelming significance are carbon, hydrogen, oxygen, nitrogen, iron, calcium, phosphorus and sulphur. Some other elements are required in much smaller amounts and a list of these is shown in the table. But even with these elements which are required in tiny quantities, the total number of elements in the body is only a small fraction of the total variety of elements to be found on earth.

Calcium	Fluorine	Magnesium	Potassium
Carbon	Hydrogen	Oxygen	Sodium
Chlorine	Iodine	Nitrogen	Sulphur
Cobalt	Iron	Phosphorus	Zinc

Table. *The main elements required by the human body*

What would happen if by the use of a 'magic knife' we took a piece of an element, say gold, and cut it into smaller and smaller pieces? Could we go on splitting the gold indefinitely or would we reach a point at which the minute piece of metal if further split up would cease to be gold? There would be such a point: the smallest particle of gold which can exist and still be gold is known as an atom of gold. We can extend the definition to all elements and say that the smallest particle of any element which retains the properties of that element is an atom.

The structure of atoms

Suppose that we could break up an atom of an element, what would be the result? We would get a number of much smaller particles, the most important of which are the protons, the neutrons and the electrons. All atoms are made up of protons, neutrons and electrons. It is the different combinations in which these fundamental particles

		Charge	Weight
(dotted circle)	Proton	+	1
(open circle)	Neutron	Nil	1
•	Electron	–	1/2000

Fig.4.1. The properties of protons, neutrons and electrons (see text).

are found which give the atoms of the elements their characteristic properties.

The atom consists of a heavy central nucleus rather like the sun, around which is a cloud of light electrons, rather like the planets of the solar system. The heavy central nucleus is made up of protons and neutrons. All protons and neutrons have the same weight which is arbitrarily stated to be about 1. They differ in that each proton carries a single unit of positive electricity while each neutron (as the

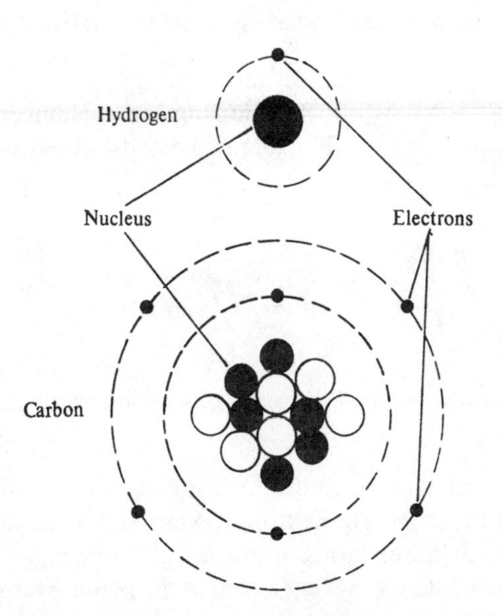

Fig.4.2. Sketches of a hydrogen atom and a carbon atom. The black circles in the nucleus are protons and the white ones are neutrons.

name implies) is electrically neutral. The electrons are quite different both from protons and neutrons. Each is only about one two-thousandth (1/2,000) of the weight of a proton or a neutron and each carries a single unit of negative electricity. About 2,000 electrons would therefore be needed to make up the same weight as a single proton but in contrast, the positive charge on a proton can be precisely neutralized by the single negative charge on an electron. When a proton and an electron come together, the positive and negative charges cancel out and the result is neutral.

In any atom the number of protons in the nucleus is equal to the number of electrons surrounding the nucleus. Atoms as a whole are therefore electrically neutral because the positive and negative electrical charges cancel out. The type of an atom (i.e. which element it is) depends on the number of protons in the nucleus. For example, all hydrogen atoms have a single proton in the nucleus and a single electron outside: conversely, any atom which has only a single proton must be an atom of hydrogen and it cannot be anything else. Carbon has six protons and an atom which has six protons must be carbon and nothing else. Calcium has twenty protons, iron has twenty-six, iodine has fifty-three, lead has eighty-two and so on. The number of protons in the nucleus is known as the atomic number. It determines what sort of atom it is and what chemical properties it will have.

	Atomic Number (Number of Protons)	Number of Electrons
Hydrogen (H)	1	1
Carbon (C)	6	6
Calcium (Ca)	20	20
Iron (Fe)	26	26
Iodine (I)	53	53
Lead	82	82

Fig.4.3. Protons and electrons in the atoms of some important substances.

The number of neutrons in the nucleus does not affect the chemical properties of the atom. The neutrons do however affect the weight of the atom. Each atom has a characteristic weight but this is so extremely small that it is not expressed in the familiar terms of grams or ounces: these units are far too large for the minute atoms. Instead, atomic weight is expressed, roughly speaking, as the ratio of the weight of an atom to the weight of an atom of hydrogen. Hydrogen consists simply of one proton, one electron and no neutrons and its atomic weight is arbitrarily taken as 1. Oxygen has eight neutrons

	Atomic Number (Protons)	Neutrons	Electrons	Atomic Weight
Hydrogen	1	0	1	1
Carbon	6	6	6	12
Oxygen	8	8	8	16

Fig.4.4. The structures of hydrogen, carbon and oxygen atoms.

and eight protons in the nucleus: the oxygen atom is therefore about sixteen times as heavy as the hydrogen atom and the atomic weight of oxygen is 16. Carbon usually has six protons and six neutrons in the nucleus: its atom is therefore twelve times as heavy as an atom of hydrogen and its atomic weight is said to be 12. The atomic weights of all the other elements are determined in the same way.

A gram-atom is said to be X grams of a substance whose atomic weight is X. For example, 1 gram-atom of hydrogen is 1 g of hydro-

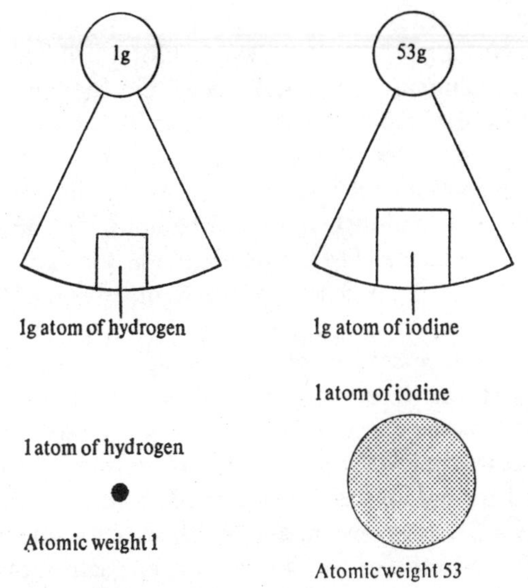

Fig.4.5. The concept of the gram atom (see text).

gen. 1 gram-atom of carbon is 12 g of carbon. 1 gram-atom of oxygen is 16 g of oxygen. 1 gram-atom of iodine is 53 g of iodine and so on. The important thing to remember about gram-atoms is that 1 gram-atom of any element contains precisely the same number of atoms as 1 gram-atom of any other element. At first this may sound odd, when you know that 1 gram-atom of hydrogen weighs 1 g while 1 gram-atom of iodine weighs 53 g. But although 1 gram-atom of iodine weighs fifty-three times as much as 1 gram-atom of hydrogen, it must be remembered that each atom of iodine weighs fifty-three times as much as 1 atom of hydrogen. Therefore the number of hydrogen atoms in 1 gram-atom of hydrogen must be equal to the number of iodine atoms in 1 gram-atom of iodine and the number of oxygen atoms in 1 gram-atom of oxygen.

Atoms and ions

Some atoms can very readily lose one or more of their electrons. Other types of atom can very readily pick up one or more extra electrons. Hydrogen, sodium and potassium are elements important in medicine which can each lose one electron. When the electron is lost the number of protons in the nucleus of the atom does not change. There is therefore one positive charge on a proton which is not balanced by a corresponding negative charge on an electron. As a result the particle has one excess unit of positive charge and it can be written H^+, Na^+ or K^+. (H is the symbol for a hydrogen atom, Na the one for a sodium atom and K the one for an atom of potassium.) Such particles are said to be 'charged' because they are not electrically neutral. These particles, derived from the loss of one or more electrons from an atom, are known no longer as atoms but as positive ions (sometimes as cations). It should be noted that the weight of an ion is virtually identical to the weight of the parent atom since the electrons are so very light in comparison to the weight of the atom as a whole.

There are two medically important examples of atoms which can lose more than one electron. Calcium, important in bone, blood clotting and nervous function, readily loses two electrons to leave an ion with two spare protons and therefore two positive charges. The chemical symbol for calcium is Ca and so the calcium ion can be written as Ca^{++}. Iron, essential for the manufacture of haemoglobin in the red cells of the blood, can lose either two or three electrons. The chemical symbol for iron is Fe and so the ions may be written as Fe^{++} or Fe^{+++}. Fe^{++} is known as ferrous iron and Fe^{+++} is known as ferric iron.

In contrast to the above examples, other elements can pick up extra electrons. These are not compensated by extra protons in the nucleus and so the result is a negatively charged ion. Medically the two most important negative ions derived from elements are those which are formed from chlorine and iodine, each of which can pick up a single extra electron. The chemical symbol for chlorine is Cl and so the chloride ion which is derived from it can be written Cl^-. The chemical symbol for iodine is I and so the iodide ion derived from it can be written I^-.

Combinations of atoms and ions

It is relatively unusual to find pure elements either in the earth in general or in the body in particular. Most commonly the elements are found in the form of mixtures or of compounds. The difference between a mixture and a compound is quite important. In a mixture, several different types of atoms may be found together but there are no links between the atoms of different elements: the elements are found as elements and their atoms are merely mixed up with other atoms without any form of chemical combination. Ordinary air is an excellent example of a mixture. About 20 per cent of it consists of the element oxygen and about 80 per cent consists of the element nitrogen. The two sorts of atoms are simply mixed up together with no firm links between them.

Compounds are quite different. In a compound the atoms of two or more elements are not simply mixed together: there are definite links between the various types of element. We saw earlier that each element consists of fundamental units known as atoms which cannot be broken down further if they are to retain their characteristic properties. Similarly, compounds consist of fundamental units known as molecules which cannot be broken down further if they are to retain their properties. For example, the important substance carbon dioxide consists of molecules each of which contains two atoms of oxygen and one of carbon. The chemical symbol for carbon is C and for oxygen is O and so carbon dioxide can be written CO_2. If carbon dioxide is broken down to the individual atoms of carbon and oxygen it loses the properties of carbon dioxide. Similarly, each molecule of water contains two atoms of hydrogen and one of oxygen: it can therefore be written H_2O. If a water molecule is split up into hydrogen and oxygen it immediately loses the properties of water. A molecule therefore is the smallest unit of a compound which can exist and which can have the characteristic chemical properties of that compound.

Valency

There are relatively simple rules which govern the ways in which atoms can combine with one another. Each type of atom is said to have a particular valency (although some atoms have more than one, but that need not concern us for the moment). The meaning of this word can best be demonstrated by examples. For instance, hydrogen, chlorine, potassium, sodium and iodine all have a valency of 1. Calcium and oxygen have a valency of 2. By this we mean that an atom which has a valency of 2 can combine with one other atom which has a valency of 2 or with two other atoms each of which has a valency of 1. Calcium has a valency of 2 and chlorine has a valency of 1: this means that two chlorine atoms can combine with one calcium atom. The resulting compound, calcium chloride, can be written $CaCl_2$. Oxygen has a valency of 2 and hydrogen has a valency

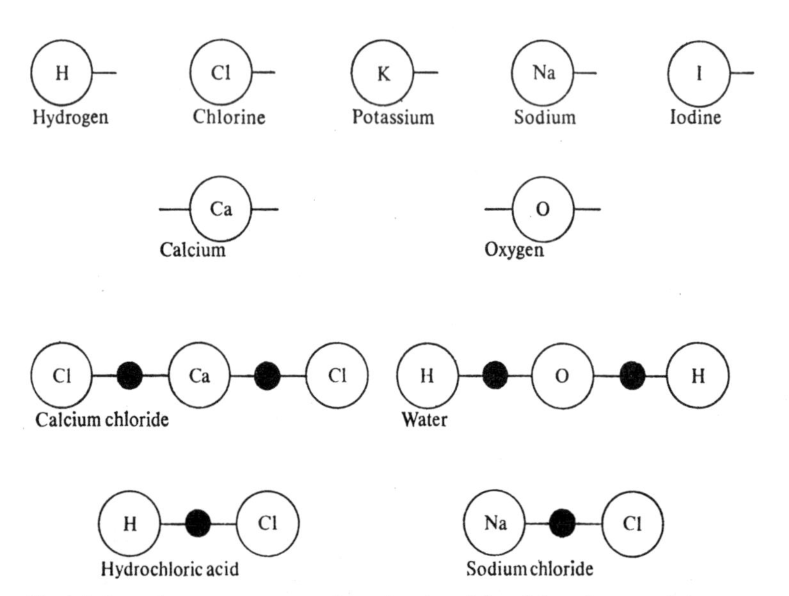

Fig.4.6. Some important atoms with valencies of 1 and 2 and some of the compounds they can form (see text).

of 1: therefore when hydrogen and oxygen combine together to form water, two hydrogen atoms combine with each oxygen atom giving the familiar chemical formula for water of H_2O. Hydrogen and chlorine both have a valency of 1: therefore when hydrogen combines with chlorine to give hydrochloric acid, one hydrogen atom combines with one chlorine atom (HCl). Sodium and chlorine also both have a valency of 1: when common salt or sodium chloride is formed, one sodium atom combines with one chlorine atom to give NaCl.

Although often regarded as a very old-fashioned way of looking at things, it is often easiest to understand the concept of valency if one imagines atoms as being little balls with hooks on them. Atoms of valency 1 have one hook, atoms of valency 2 have two

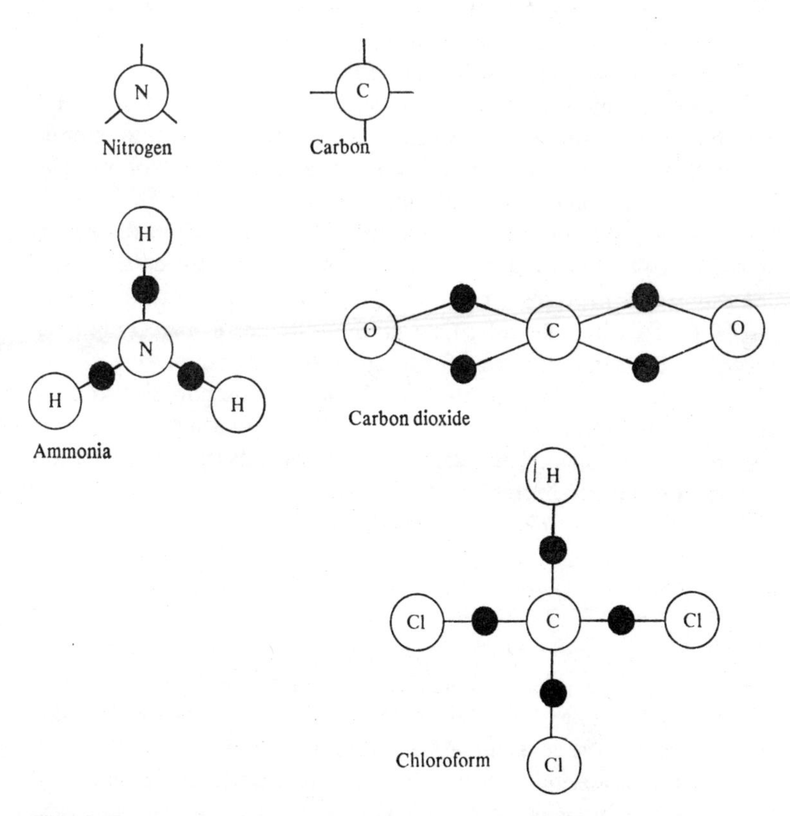

Fig.4.7. The valencies of nitrogen and carbon and some of the typical compounds they form.

hooks, atoms of valency 3 have three hooks and so on. In this way it is easy to appreciate the concept of valency.

Nitrogen is an important substance in medicine which has a valency of 3. It can, for example, combine with three hydrogen atoms to give ammonia (NH_3) a major and toxic product of the breakdown of proteins in the body. Carbon is an element which has a valency of 4. Each carbon atom can combine with four atoms which have a valency of 1, or with two which have a valency of 2. Carbon dioxide (CO_2) and chloroform ($CHCl_3$) are typical carbon compounds.

Iron is unusual in that in the ferrous state it has a valency of 2: ferrous chloride therefore contains two chlorine atoms and can be written $FeCl_2$. In the ferric state it has a valency of 3: therefore ferric chloride has three chlorine atoms and can be written $FeCl_3$. Iron thus has two alternative valencies.

Earlier in this chapter we discussed the concept of the gram-atom. The concept of the gram-molecule is very closely related. The weight of a molecule is defined as the sum of the weights of the atoms which make up the molecule. For instance, carbon dioxide consists of 1 atom of carbon and two of oxygen. Each carbon atom has an atomic weight of 12 and each oxygen atom an atomic weight of 16. The weight of a carbon dioxide molecule is therefore equal to $12+(2\times16)$ which is 44. We therefore say that the molecular weight of carbon dioxide is 44. One gram-molecule is defined as the molecular weight of a substance in grams. Therefore 1 gram-molecule of carbon dioxide is 44 g of carbon dioxide. One gram-molecule of any compound contains the same number of molecules as 1 gram-molecule of any other compound. One gram-molecule of any compound also contains the same number of molecules as there are atoms in 1 gram-atom of any element. The reasoning is the same as that set out in the section on the structure of atoms.

Covalent and ionic compounds

There are two main types of valency link between the elements. In one type, the valency 'hooks' become firmly attached together, closely binding one atom to the next. Compounds made by links like this are said to be covalent. An important characteristic of covalent compounds, especially when their molecules are large, is that they often dissolve more easily in liquids like alcohol or acetone than they do in water.

Ionic compounds are quite different. As their name suggests they are really compounds made up of ions rather than of atoms. The links between the ions are not firm but they depend on the fact that positive and negative units of electricity tend to be attracted to one another. In contrast, positive charges tend to repel other charges and negative charges tend to repel other negative charges. Ionic compounds therefore consist of positive and negative ions bound together by mutual attraction: the two types of ion are in such proportions that the two types of charge cancel one another out and the resulting compound is electrically neutral. Atoms of valency 1 form ions with one charge, those of valency 2 form ions with two charges and so on. Thus sodium, potassium, hydrogen and chlorine are all of valency 1 and their ions can be written Na^+, K^+, H^+ and Cl^-. Sodium, potassium and hydrogen ions are all positively charged. They therefore repel one another and they cannot form compounds with one another. But each of the three is attracted to the negatively charged chloride ions and so each can combine with chloride to form a compound. The three compounds can be written Na^+Cl^- (sodium chloride or common salt), K^+Cl^- (potassium chloride) and H^+Cl^- (hydrochloric acid). In each case one ion with a single positive charge combines with one ion with a single negative charge.

Calcium is a good example of a biologically important element which can form an ion with two positive charges as a result of the loss of two electrons. Each calcium ion therefore needs two ions carrying a single negative charge to neutralize it. For example one calcium ion will link with two chloride ions to give calcium chloride. This can be written $Ca^{++}Cl^-{}_2$. (The 2 below the line indicates that two chloride ions are present.)

Some important ions are not derived from single elements. Instead they are formed from groups of atoms which are combined together and which then, as a group, gain an electron or a proton. The gaining of a proton is another example of the formation of an ion which we have not yet discussed: it is much less common than the loss or gain of electrons. An atom which gains a proton obviously gains an extra positive-unit of electricity and therefore becomes positively charged. The ammonium ion is the best example of a positively charged ion made from a group of atoms. The ammonia molecule (NH_3) can combine with a hydrogen ion (H^+) to give an ammonium ion ($NH_4{}^+$).

There are many more negatively charged ions formed from groups of atoms. The most important are described below. In these cases each ionic group may be said to have a valency equal to the number of charges on the ion. Sulphate, for example, is an ion which has a valency of 2. It is negatively charged and therefore it can combine with two positive ions each carrying a single charge. Typical compounds in which the sulphate ion is of importance are sulphuric acid ($H^+{}_2SO_4{}^{--}$) and sodium sulphate ($Na^+{}_2SO_4{}^{--}$). The bicarbonate ion is of valency 1 and can combine with one positively charged ion such as the ions of sodium and potassium to give sodium bicarbonate ($Na^+HCO_3{}^-$) or potassium bicarbonate ($K^+HCO_3{}^-$). The ammonium ion is also of valency 1 and can combine with one negative ion with a single negative charge such as chloride to give ammonium chloride ($NH_4{}^+Cl^-$) and so on.

Gram equivalents

One gram-atom is X gram of any element whose atomic weight is X. One gram-ion is X gram of the ions formed from any element whose atomic weight is X. There is no effective difference between one gram-atom of an element and one gram-ion of the same element because the electrons which are lost or gained are so minute in comparison to the size of the atom. With complex ions the numerical value for the ionic weight can easily be calculated by adding together the atomic weights of the atoms which make up the ion. The bicarbonate ion, for example, contains one hydrogen, one carbon and three oxygen atoms. The atomic weight of hydrogen is 1, of carbon is 12 and of oxygen is 16. The total weight of the ion is therefore $1+12+(3\times16)$ and is 61. One gram-ion of bicarbonate is therefore 61 g of bicarbonate.

As we saw earlier, 1 gram-atom or -ion of any element must contain the same number of atoms or ions as 1 gram-atom or -ion of any other element. Thus, 1 gram-atom of sodium contains precisely as many atoms of sodium as 1 gram-atom of chlorine contains atoms of chlorine. Both sodium and chlorine are of valency 1 and this means that one atom of sodium can combine with one atom of chlorine. The atomic weight of sodium is 23 and so 1 gram-atom of sodium is 23 g of sodium. The atomic weight of chlorine is 35·5 and so 1 gram-atom of chlorine is 35·5 g. Therefore when 23 g of sodium combine with 35·5 g of chlorine we can be sure that each sodium atom will be able to combine with a chlorine atom and that there will be no sodium or chlorine left over at the end.

What happens when an element of valency 1 combines with an

element of valency 2, say chlorine, with calcium. One gram-atom of calcium contains twice as many valency bonds as it does atoms since each atom has two valency bonds. Therefore 1 gram-atom of calcium will combine with 2 gram-atoms of chlorine.

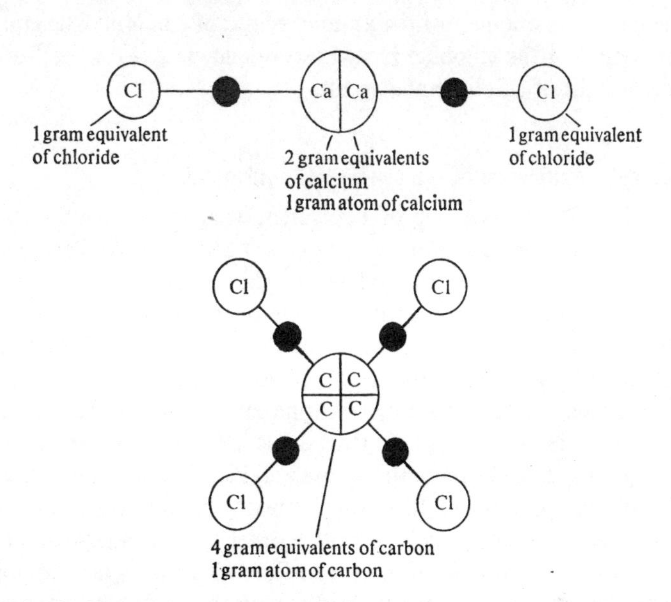

Fig.4.8. 1 gram-equivalent of a substance is equal to 1 gram-atom or 1 gram-ion divided by the valency of that atom or ion.

This is where the concept of the gram-equivalent comes in. It is clear that since 1 gram-atom of calcium will combine with 2 gram-atoms of chlorine, then $\frac{1}{2}$ gram-atom of calcium will combine with 1 gram-atom of chlorine. Carbon has four valency bonds on each atom and therefore 1 gram-atom of carbon will combine with 4 gram-atoms of chlorine. One-quarter of a gram-atom of carbon will therefore combine with 1 gram-atom of chlorine.

The equivalent weight of any substance is defined as the weight of that substance which will combine with 1 gram-atom of a substance like hydrogen or chlorine which has a valency of 1. Since 1 gram-atom of any univalent substance will combine with 1 gram-atom of any other substance with a valency of 1, in the case of atoms and ions whose valency is 1, the atomic (or ionic) weights and the equivalent weights are the same. This is not true of atoms or ions whose valency is greater than 1. By the reasoning demonstrated in the previous paragraph it is possible to see that the equivalent weight is equal to the atomic or the ionic weight divided by the valency. Thus the equivalent weight of carbon whose valency is 4 is one-quarter of the atomic weight of carbon. The equivalent weight of calcium whose valency is 2 is one-half of the atomic weight of calcium. The equivalent weight of the sulphate ion whose valency is 2 is one-half of the ionic weight of sulphate and so on.

Medical significance of the concept of equivalents

Why should the concept of equivalent weight be so important? It is because one equivalent weight of any substance will react with (is equivalent to) one equivalent weight of any other. The use of equivalents makes the understanding of many chemical problems much easier. For example, if someone said to you that a litre of fluid contained 61 g of bicarbonate and 23 g of sodium, the meaning of this statement would not be very apparent unless you happened to know that the ionic weight of bicarbonate is 61 and the atomic weight of sodium is 23 and that both sodium and bicarbonate have a valency of 1. If you did know all these things then you would also know that there was just enough sodium in the water to combine with all the bicarbonate without leaving any sodium or bicarbonate left over. On the other hand if someone said to you that a litre of water contained one equivalent of bicarbonate and one equivalent of sodium, then the significance of the statement would be immediately apparent to you without the need for any further knowledge or calculation. The wider significance of this will become apparent in the next chapter in the section on the neutrality of solutions.

One equivalent is the equivalent weight in grams of any element or ion. It tends to be rather a large quantity. It is therefore often much more convenient to use the term milliequivalent instead. One milliequivalent is one-thousandth of one equivalent. For instance, as a nurse you will frequently read laboratory reports of estimations of the sodium and potassium concentrations in the plasma of patients. These concentrations are usually expressed in terms of milliequivalents per litre. The normal concentration of sodium is usually 140 milliequivalents/litre while the normal concentration of potassium is about 4·5 milliequivalents/litre. Since both sodium and potassium are of valency 1, 1 milliequivalent of sodium contains the same number of atoms as 1 milliequivalent of potassium. Therefore by knowing the concentrations in milliequivalents one can say instantaneously that there are about thirty times as many sodium ions in plasma as there are potassium ions.

Organic and inorganic compounds

Chemists often divide chemical compounds up into those which are inorganic and those which are organic. Originally the word inorganic was used to describe substances, such as salt, which could be found in the non-living world. In contrast the word organic was used to describe those compounds which could only be found in living creatures. All living creatures of course contain both organic and inorganic substances.

All the substances characteristically found only in living things contain large amounts of carbon. Many similar substances can now

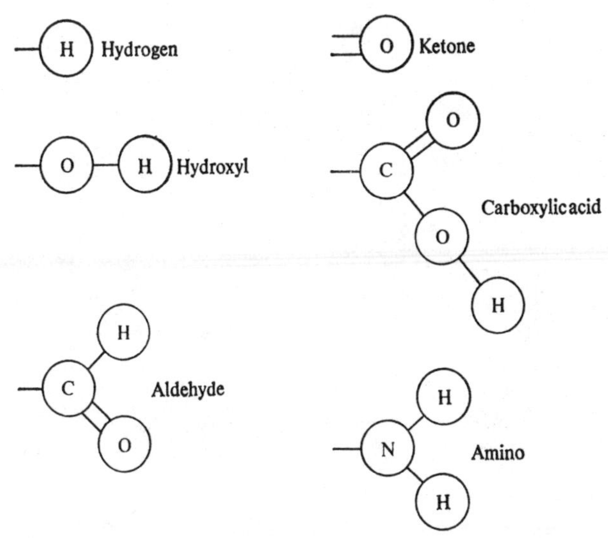

Fig.4.9. Some important groups of atoms found in organic substances.

be manufactured artificially in the laboratory and so organic chemistry today has come to mean the chemistry of complex carbon compounds. Because it has a valency of 4, carbon is a very versatile substance which can form many complex structures. Many organic compounds are based on chains of carbon atoms or on rings containing five or six carbon atoms. Attached to these chains and rings there may be a wide variety of other atoms or groups of atoms but the most important ones are all shown in figure 4.9. They are hydrogen (H), hydroxyl (OH) aldehyde (CHO), ketone (O), carboxylic acid (COOH) and amino (NH_2). There is no need for you to memorize the names or the structures of these groups. They have been described here so that if they are mentioned you can look them up to see what they consist of.

Of all these groups, the carboxylic acid and amino ones are particularly interesting. They occur in large numbers in proteins and many of the peculiar properties of proteins depend on their presence. The carboxylic acid group can split to give the ionic structures $-COO^-$ and H^+. The amino group can pick up an extra proton to become $-NH_3^+$. In this way the amino group is similar to the ammonia molecule.

5

Isotopes and radioactivity

As we have seen, the number of protons in the nucleus, the atomic number, determines the chemical properties of an atom. The number of protons is fixed and immutable for any particular element. But the number of neutrons in the nucleus is not fixed. Atoms of one element which therefore all have the same number of protons in the nucleus may have different numbers of neutrons. Since each neutron weighs as much as each proton, two atoms of the same element which have the same number of protons, but different numbers of neutrons, will have significantly different atomic weights. Atoms of the same element but which have different weights because of differing numbers of neutrons are known as isotopes of that element. This concept may be a little difficult to understand when expressed in abstract terms but some examples should make it clear.

Carbon normally has six protons and six neutrons in the nucleus and so the atomic weight of carbon is normally 12. But some carbon atoms have six protons and eight neutrons and therefore their atomic weight is 14. Normal carbon atoms may be written as ^{12}C or carbon-12 while the abnormally heavy ones are known as ^{14}C or carbon-14. Carbon-12 and carbon-14 are therefore two isotopes of carbon: they both have the same chemical properties but they have different atomic weights. Iodine atoms normally have fifty-three protons and seventy-four neutrons giving an atomic weight of 127. But some iodine atoms have seventy-eight neutrons making a total weight of 131 (^{131}I or iodine-131). Others may have seventy-nine neutrons making a total weight of 132 (^{132}I or iodine-132).

Many isotopes are naturally occurring although natural preparations of any element usually contain an overwhelming preponderance of one particular isotope. Natural carbon contains only a minute proportion of carbon-14 and so the average atomic weight of naturally occurring carbon atoms is very close to the atomic weight of carbon-12. Chlorine is an important exception. About one-quarter of naturally occurring chlorine atoms contain seventeen protons and

	Protons	Neutrons	Atomic Weight
Carbon-12	6	6	12
Carbon-14	6	8	14
Iodine-127	53	74	127
Iodine-131	53	78	131
Iodine-132	53	79	132
Chlorine-35	17	18	35
Chlorine-37	17	20	37

Fig.5.1. The structures of some important isotopes.

twenty neutrons giving an atomic weight of 37. The remaining three-quarters of the atoms contain seventeen protons and eighteen neutrons giving an atomic weight of 35. Natural chlorine therefore has an average atomic weight of 35·5.

Radioactive isotopes

Many isotopes are stable and the atoms show no tendency to break up into their constituent protons, neutrons and electrons. But this is not true of all isotopes. In some cases the particular combination of protons and neutrons tends to be unstable and the atoms spontaneously break up into smaller components, giving off various types of radiation in the process. Such unstable isotopes are said to be radioactive. A few radioactive isotopes occur naturally but most are now made artificially in atomic reactors.

Three main types of radiation are emitted by radioactive substances. These have been given the names of Greek letters, alpha, beta and gamma. Alpha radiation consists of particles, each one of which contains two protons and two neutrons. Alpha radiation is very easily stopped on impact with anything, and even in air it cannot travel for more than a few centimetres. Beta radiation consists of electrons: it has greater penetrating power than alpha radiation but because the electrons are so light it is easily deviated from a straight line. Gamma radiation consists of waves of the same general sort as those in radio, light and X-rays but it is of very high penetrating power and can easily pass through the human body.

Obviously when an atom in a sample of radioactive material disintegrates, that atom has gone and there is less potential radioactivity in the sample. When an atom disintegrates it is often said

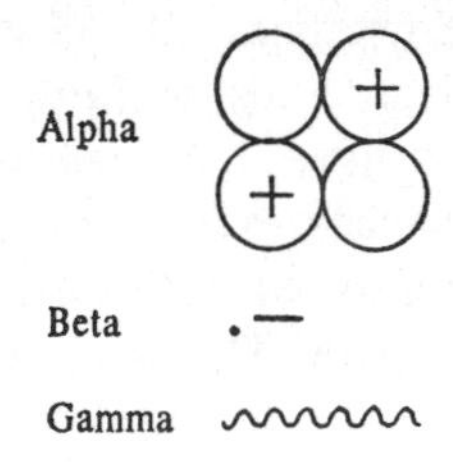

Fig.5.2. The main types of radiation (see text).

to have decayed. Isotopes differ enormously in their stability and therefore in their rate of decay. The stability of radioactive isotopes is usually expressed in terms of the half-life. The half-life is the time taken for half the atoms in a sample of the isotope to disintegrate. Some isotopes are so unstable that their half-lives are measured in fractions of a second while others are so stable that their half-lives are measured in thousands of years.

Radioactive isotopes in medicine

Radioactivity and the use of isotopes are becoming more and more important in medical science. The great majority of applications are to be found in research laboratories and are not of practical value in the diagnosis and treatment of ordinary patients. But there are at least two highly practical uses of radioactivity which are widespread throughout the world and it seems probable that in the near future more and more practical uses will be found.

One important use at present depends on the fact that radiation damages and kills cells. It does this in ways which have not yet been clearly worked out. However, it is known for certain that the main effect of radiation is on cells which are rapidly dividing and multiplying. These cells can be killed by low doses of radiation which have little apparent effect on non-dividing cells.

A number of cell types in the body normally divide and multiply rapidly. The surface of the skin and the lining of the gut are continually being damaged and renewed. The red and white cells of the blood also have relatively short lives and must be replaced by the continual activity of the bone marrow. Cancer cells are abnormal cells which divide and multiply rapidly and it has been found that they are very susceptible to destruction by radiation. If a malignant tumour is bombarded with a sufficient dose of radiation its growth is almost always slowed down. Sometimes all the cancerous cells may be completely destroyed, giving a real and permanent cure. Most large hospitals, therefore, now have machines for generating beams of high-intensity radiation which can then be used to bombard tumours. It is important, of course, to reduce the exposure of normal tissues to such beams to a minimum. The skin, gut and bone marrow in particular may be permanently damaged by such exposure.

Unfortunately radiation is by no means an unmixed blessing. While high doses can kill tumour cells, even quite low doses can cause other normal tissues to change into malignant cancers. This risk even applies to X-rays although these are not so dangerous as other types of radiation. Nevertheless some types of malignant disease, such as leukaemia, are significantly commoner in radiologists than in other members of the medical profession. It is therefore important to avoid any unnecessary exposure to radiation. And that applies to the X-ray department as well as to the cancer radiotherapy unit.

Another important medical use of radioactivity is in the diagnosis of diseases of the thyroid gland. The thyroid hormone which is manufactured by the gland contains iodine, and without iodine the hormone cannot be synthesized. If iodine is injected into the blood, it is taken up by the cells of the thyroid. If the gland is working too vigorously, the iodine is removed from the blood abnormally quickly. If it is working too slowly, the iodine is removed from the blood abnormally slowly.

Radioactive iodine is treated by the thyroid in just the same way as ordinary iodine. Thus a dose of radioactive iodine given either by mouth or by injection is removed from the blood at the same rate as is ordinary iodine. Therefore the rate at which the radioactive iodine accumulates in the thyroid gland gives a very good measure of the state of thyroid activity. The beauty of the radioactivity is that it can very easily be detected by a counter placed over the gland on the neck. The radiation easily penetrates the skin and can be picked up by the counting device. The rate at which the radioactive iodine enters the gland can therefore be easily detected and compared with the known normal rate. This is therefore a rapid and accurate method of deciding whether the thyroid gland is normal or is over- or under-active.

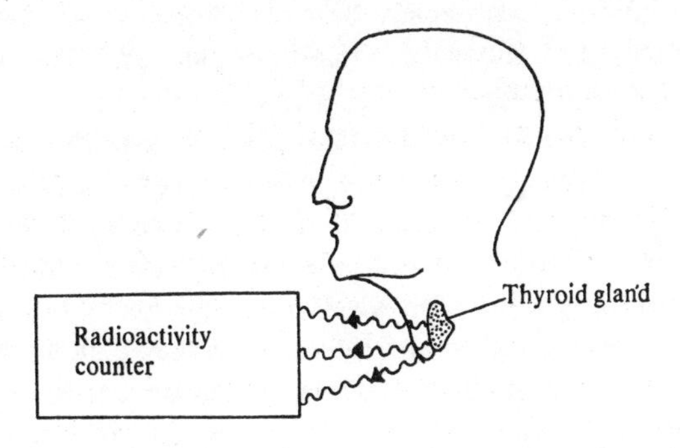

Fig.5.3. The use of radioactive iodine to study the activity of the thyroid gland (see text).

The thyroid tests using radioactive iodine also illustrate the importance of knowing the half-life of an isotope. For many years iodine-131 which has a half-life of about 8 days has been used. This means that the patient is exposed to radiation, even though at a very low level, for many days after the test is usefully completed. More recently therefore, iodine-132 has begun to be used. This has a half-life of only 2·3 hours. Although it is rather less accurate than iodine-131 it does enable the amount of exposure of the patient to radiation to be reduced.

X-rays

X-rays (sometimes known as Röntgen rays) are similar to gamma rays but they have a rather longer wavelength and are neither so penetrating nor so dangerous. They have two important properties which make them of immense value in medicine. First they can be detected easily when they fall on plates which have much the same composition as ordinary photographic film. When a lot of X-radiation falls on a particular area of an X-ray film that area appears dark, while when little radiation falls on an area, that area appears to be light in colour.

The second important point about X-rays is that they can be relatively effectively stopped by the heavy atoms of certain elements of which calcium, barium and iodine are probably the most important medically. The bones are, of course, very rich in calcium. Therefore when an X-ray is taken of any part of the body, the bones stand out as light areas against a dark background and any deformities or breaks in the bones can easily be seen. Stones which form in the kidneys or urinary tract are also rich in calcium and they can usually be clearly seen on X-rays.

Barium is important in the taking of X-ray films of the gut. Normally X-rays easily pass through the soft tissues of the abdomen and very little can be seen of soft-tissue structure on the ordinary X-ray film. But if compounds containing barium are mixed up into a sort of porridge and then swallowed, the heavy barium atoms clearly outline the stomach and intestines as they pass down. The X-rays cannot pass through the barium and so the barium-filled areas appear as opaque white patches. The lower end of the gut can be clearly outlined by administering the barium compounds in the form of an enema.

Iodine too is radio-opaque: it does not let X-rays through. Some iodine compounds are specifically excreted by the liver into the bile and they can be used to take good X-ray pictures outlining the gall bladder and bile ducts. Others are excreted in the urine and can be used to outline the kidneys, ureters and bladder. Using these techniques abnormalities of the biliary and urinary tracts can often be easily identified.

6

Solutions

Between 50 and 60 per cent of the total mass of the human body consists of water. Much of the remainder consists of substances dissolved in water either in cells, in the extracellular fluid or in the blood. When a solid is dissolved in a liquid, the result is known as a solution. The liquid part of the solution is sometimes known as the solvent and the dissolved solid is known as the solute. Because of the immense amount of water in the body it is very important to understand something of the behaviour of solutions when the solvent is water.

The nature of a solution

A solution is said to exist when a solid is mixed with a liquid and when the particles of the solid material are dispersed throughout the liquid so that they have no tendency to settle out and cannot be seen either with the naked eye or with the most powerful of microscopes. In fact, in a solution all the individual molecules, atoms or ions, of which the solid is made float freely and independently of one another. A suspension is quite different. A suspension exists when a solid is mixed with a liquid but when the solid particles can be easily seen and tend to settle down to the bottom.

The difference between a solution and a suspension can be easily demonstrated by putting a spoonful of sugar into one bottle of water and a spoonful of ground glass particles into another bottle of water. What happens when the two bottles are shaken vigorously? The sugar particles disappear completely: they cannot be seen and even if the jar is left sealed and undisturbed for a whole year or longer the sugar will not settle out. The sugar molecules will remain dispersed throughout the water forming a true solution. During vigorous shaking, the ground glass particles will also become evenly distributed throughout the water. But the glass particles can always be clearly seen, even by the naked eye, and once the stirring stops they will quickly settle down to the bottom of the bottle.

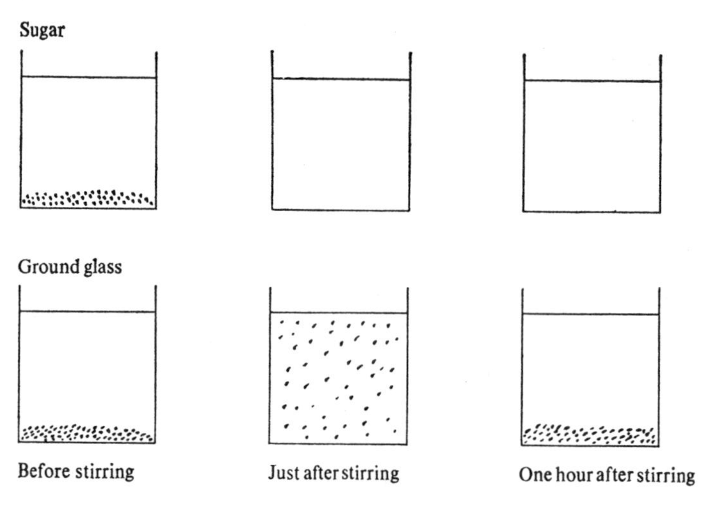

Fig.6.1. The difference between a suspension and a solution.

Blood itself also illustrates the distinction between a suspension and a solution. It contains many substances truly dissolved in it, such as salt, proteins and glucose. No matter how long blood is left to stand, these substances will remain dissolved and will not settle out. But blood also contains suspended in it the red and white cells which are kept evenly distributed by the continual circulation and mixing of the blood. But if blood is prevented from clotting by adding an anti-clotting agent and is then allowed to stand, the red and white cells slowly settle out leaving a clear fluid known as the plasma above them.

Sedimentation rate

The settling out of the cells of the blood is the basis of a simple and very useful ward test known as the erythrocyte sedimentation rate (ESR) or as the blood sedimentation rate (BSR). In this test, a sample of blood is taken from a patient and treated with a chemical to stop it clotting. The blood is then sucked up into a narrow graduated tube (usually 20 cm long), mounted in a rack and left to stand for 1 hour. During this time the red cells settle down leaving above them a layer of clear plasma whose depth can easily be estimated by reading off the graduated tube. Normally the layer of

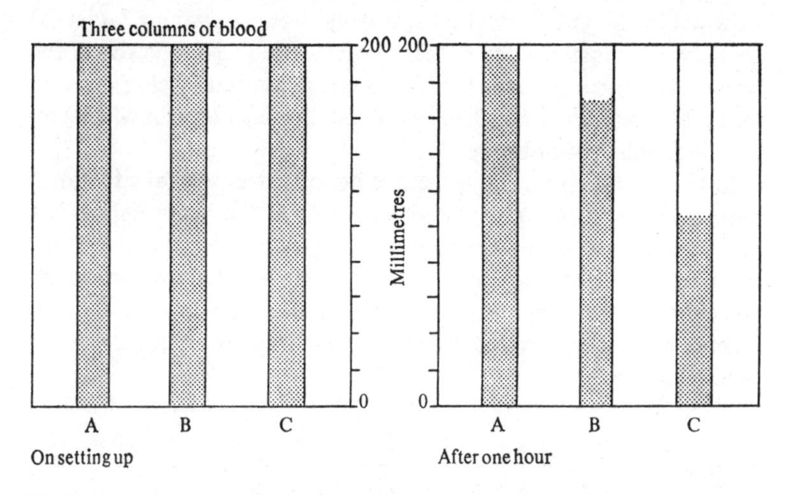

Fig.6.2. The determination of the ESR. Patient A is normal, B has a slightly raised ESR and C has a greatly raised ESR (see text).

plasma is less than 10 mm deep and the ESR is said to be less than 10 mm/hour. But all sorts of conditions, in particular infections, inflammatory diseases like rheumatic fever or rheumatoid arthritis, and cancer can greatly increase the ESR sometimes to over 100 mm/hour. In these conditions there are abnormal proteins in the blood which make the red cells stick together in large clumps. The clumps of cells settle out much more rapidly than the single separate cells in normal blood.

The ESR is a very useful screening test for the 'difficult' patient who comes to the doctor with rather vague and ill-defined symptoms. The problem is always 'Does this person have something seriously wrong or not?' If the ESR is normal it does not entirely exclude the possibility of serious disease but it does make it unlikely. On the other hand, if the ESR is raised above normal, the doctor must not rest until the cause has been found. No patient can fake a high ESR.

Emulsions

A special sort of suspension is sometimes important in medicine. This is the emulsion in which minute droplets of fatty material are evenly dispersed through water. Some emulsions in which the fat droplets are extremely small are remarkably stable and separate very slowly if at all. Other emulsions, usually those containing larger fat particles, are unstable and the fat particles rise to float on top of the water when mixing is stopped. Salad cream and many cosmetic and medicinal creams are excellent examples of emulsions with which you will be familiar in ordinary life.

In human function, emulsions are important in several situations, most notably in the transport of fats around the body. Before fats can be absorbed across the wall of the intestine into the body they must be converted into an emulsion whose particles are very small. Once the droplets enter the blood they are then carried around in the form of an emulsion to the liver where the droplets are removed and processed further.

Solubility

If you take a spoonful of salt and stir it up in a beaker of water, the salt dissolves very readily: it is said to be highly soluble. If in contrast you take a spoonful of chalk and stir it up, very little of the chalk dissolves in the water: the chalk is said to be a relatively insoluble substance.

If you take the beaker of salt solution and then continue adding spoonsful of salt, stirring after each addition, you will eventually reach a point where the salt you have added fails to dissolve. No more salt can dissolve in the water and the solution is said to be saturated with salt. When extra salt is added to a saturated solution it just fails to dissolve and remains as a sediment on the bottom.

In a saturated solution, the water is carrying the maximum amount of solid that that volume of water can hold. What happens therefore if we take a solution which is just saturated and has no sediment on the bottom and then we boil it? Boiling the solution drives off the water and reduces the volume of the solution. But boiling does not make the salt dissolved in the solution evaporate. Therefore the same amount of salt is present in a smaller volume of water. The smaller volume of water cannot carry all this salt and so some of the salt is deposited as solid salt: it is said to precipitate out. The precipitating out of dissolved solids is the cause of gall stones and of stones in the urinary tract.

Electrolytes

Many substances when they dissolve in water do not exist in the form of uncharged atoms or molecules. Instead they form ions which carry an electrical charge. For example, common salt or sodium chloride, when dissolved in water exists in the form of positively charged sodium ions (Na^+) and negatively charged chloride ions (Cl^-). Proteins also form ions: their carboxylic acid groups ($-COOH$) give off a hydrogen ion (H^+) to leave a negatively charged carboxylate ion ($-COO^-$) behind. The amino groups of proteins ($-NH_2$) pick up a hydrogen ion to give $-NH_3{}^+$. All substances which when dissolved in water form electrically charged ions in this way are said to be electrolytes. Most electrolytes are either inorganic substances or organic substances with carboxylic acid and amino groups.

It has been found that in any solution the number of positive charges of electricity on ions must equal the number of negative charges of electricity, i.e. the solution as a whole must be electrically neutral. Now you may remember that one equivalent weight of any ion must contain the same number of units of electrical charge as one equivalent weight of any other ion. This means that in any solution, the number of equivalents of positively charged ions must be precisely the same as the number of equivalents of negatively charged ions. Only then will the positive and negative charges cancel out.

This equality of positive and negative charges applies to blood just as it does to any other solution. The normal concentrations of the substances in plasma which form ions are shown in figure 6.3. The number of milliequivalents of positively charged ions precisely equals the number of milliequivalents of negatively charged ions. This enables a useful quick check to be made of any chemical analysis of the composition of blood. If the analysis is accurate the sum of the concentrations of positive ions measured in milliequivalents must be equal to the sum of the concentrations of the negative ions measured in milliequivalents. If the two are not equal the analysis must be inaccurate. If concentrations were quoted in grams per litre instead of in milliequivalents, no such easy check would be possible.

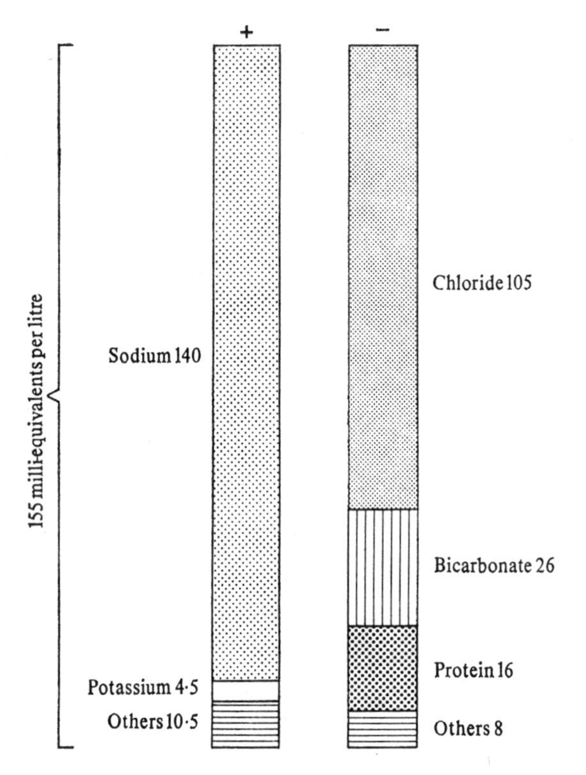

Fig.6.3. The ionic composition of plasma. The number of negative ions in milli-equivalents per litre is precisely the same as the number of positive ions.

Types of solute

As mentioned earlier, a solute is a solid substance which is dissolved in a liquid. There are many varieties of solute in the body which differ greatly in the ease with which they will dissolve in water (in other words in their solubility in water). The most important solutes are as follows:

1. Inorganic substances which are readily soluble in water such as sodium, potassium, chloride and bicarbonate ions.

2. Inorganic substances which can dissolve in water but cannot do so very easily. The most important of these are iron, calcium and phosphate ions. Iron is transported in the blood by a special mechanism so avoiding the need for it to dissolve in the plasma. It is carried by attaching it to one of the soluble plasma proteins which is

sometimes known as transferrin or iron-binding globulin. Calcium is partly bound to a plasma protein but is also partly found as free calcium ions. At their normal concentrations, both calcium and phosphate ions can usually dissolve in the plasma. But if the concentration of either calcium or phosphate increases, the plasma may then not be able to carry all the calcium and phosphate ions and the excess precipitates out in the form of solid calcium phosphate. Much of this solid material is deposited in the walls of blood vessels and in particular in the arteries where it can cause considerable damage.

The commonest disease which causes the deposition of solid calcium phosphate is renal failure. In this condition the kidneys cannot get rid of phosphate normally. The phosphate concentration in the blood therefore rises and when it reaches a certain critical level calcium phosphate is deposited from the plasma. A much rarer cause of calcium phosphate precipitation is the disease known as hyperparathyroidism. In this condition too much parathyroid hormone is produced by the parathyroid glands. This hormone acts on the bones and makes calcium pass from the bones into the plasma. Again when a critical concentration of calcium is exceeded, calcium phosphate deposits are formed.

The poor solubility of calcium can also cause problems in the urine. If the calcium concentration in the urine becomes too high, solid material may precipitate out as renal stones. These may block the urinary tract at some point causing pain and in the worst, prolonged cases, renal failure. Such stones tend to be much commoner in hot countries where water is very scarce. The minimum of water must be excreted in the urine which therefore becomes highly concentrated: this is then an ideal situation for calcium to precipitate out.

3. Small organic molecules such as those of glucose and the amino acids. These can easily dissolve in the plasma and are readily carried around the body in this way.

4. Large organic molecules. These tend to be insoluble unless, like the proteins, they form ions with electrical charges. Even large molecules which can form ions can dissolve in water. Molecules which do not form ions but which are very large are usually insoluble.

5. Fatty substances. These tend to be insoluble in water and they therefore precipitate out from body fluids rather easily. Perhaps the most important fatty substance from this point of view is the one known as cholesterol. On a diet rich in fat the cholesterol content of the plasma may become very high and much cholesterol may be

deposited in the walls of arteries giving the condition known as atheroma. Atheroma seems to be very important as a cause of both coronary heart disease and cerebrovascular accidents or strokes. Sometimes the fatty deposits themselves may completely block small arteries. At other times the uneven surfaces of the fat deposits lead to abnormal blood coagulation and so the vessel becomes blocked by a blood clot.

Cholesterol, together with bilirubin, a breakdown product of the red pigment in red cells (haemoglobin), is also important in the bile. Bile is secreted by the liver and contains large amounts of both cholesterol and bilirubin. The secreted bile is then stored in the gall bladder and while it is there water is removed from it by the wall of the bladder. This makes the bile more concentrated and as a result cholesterol and bilirubin may precipitate out as the solid masses known as gall stones. Gall stones may block the bile duct and make it impossible for the bile to be excreted. When this happens the yellow bilirubin accumulates in the blood giving the condition known as jaundice. Since in many patients the bile and pancreatic ducts enter the intestine together, gall stones may also block the pancreatic duct giving the dangerous disease known as pancreatitis. In this condition the pancreatic enzymes which normally digest food cannot escape into the gut: instead they may digest the pancreas itself together with the surrounding tissues.

I hope that it is now apparent that foı the nurse an elementary knowledge of the properties of solutions is not useless but is very important in the understanding of many different disease processes.

7

The importance of osmotic pressure

Osmotic pressure is one of the most important chemical concepts used in medicine. Despite this it is a topic which is very frequently misunderstood by both nurses and medical students. It is perhaps best to begin this discussion by describing an extremely simple experiment.

Suppose that we take two beakers full of water. In each of these beakers we immerse a glass cylinder across the bottom end of which has been tied a membrane. This membrane, like many living membranes, has the property that it will let water pass through it freely but it will not let pass the large molecules of dissolved protein. Such membranes which are permeable to a liquid but not to a solid dissolved in that liquid are often known as semi-permeable.

Suppose then that we put inside the glass cylinder a solution of a protein containing 5 g of protein per litre and that in the other glass cylinder we put a solution of the same protein but containing 10 g of protein per litre. We adjust the levels of the water in the beakers and the protein solution in the cylinders so that they are just the same. We then leave the experiment and do not look at it again for 24 hours.

What then do we find? The situation has certainly changed considerably. The levels of the protein solution in the cylinders are no longer the same as the levels of the water in the beakers. Water has been sucked across the membrane from the water in the beaker into the cylinder and the fluid level in the cylinders is now much higher than that in the beakers. Moreover, the water level in the cylinder which had a protein solution containing 10 g/litre is just twice as high as that in the cylinder which had a protein solution containing 5 g/litre.

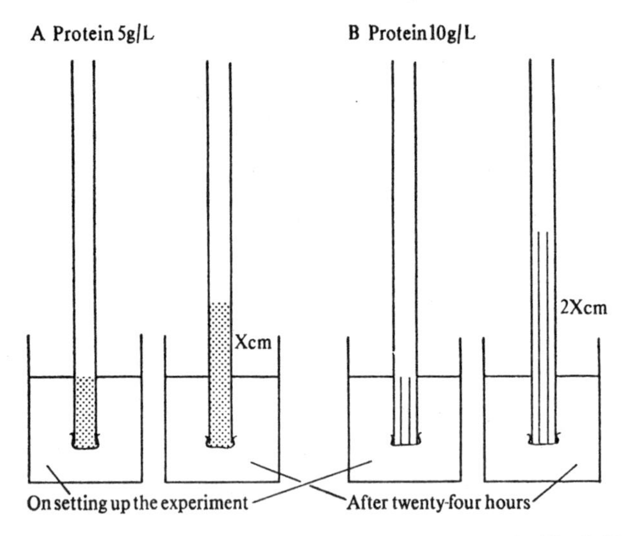

Fig.7.1. Experiment to demonstrate osmotic pressure (see text). The fluid in the beakers is pure water.

Explanation of the experiment

What seems to have happened is that water has been drawn from the pure water into the protein solution. This movement of water has raised the fluid level in the cylinder above that in the beaker. The solution in the cylinders is exerting a pressure on the membrane equivalent to X cm of water in the case of the dilute solution and 2X cm of water in the case of the strong solution. This column of liquid is therefore pressing down on the membrane and if the membrane were punctured, the solution would flow out of the cylinder into the beaker until the two levels were again equal. What seems to happen is that so long as the membrane is intact water is pulled into the cylinder: this process continues until the pressure of the column of solution is just sufficient to oppose the force pulling water into the cylinder. When the force of the column of fluid pressing down is just equal to the force of water movement pressing up then no further movement can take place. The force which drags water from the beaker into the protein solution is known as an osmotic force. The pressure which this force sets up can be measured by noting the difference in fluid levels between the beaker and the cylinder.

The fact that the osmotic pressure is doubled when the concentration of protein is doubled, tells us something else about osmotic forces. This is that the size of the osmotic force depends on the concentration of the solution exerting that force. The more concentrated the solution thc greater is the osmotic pressure set up.

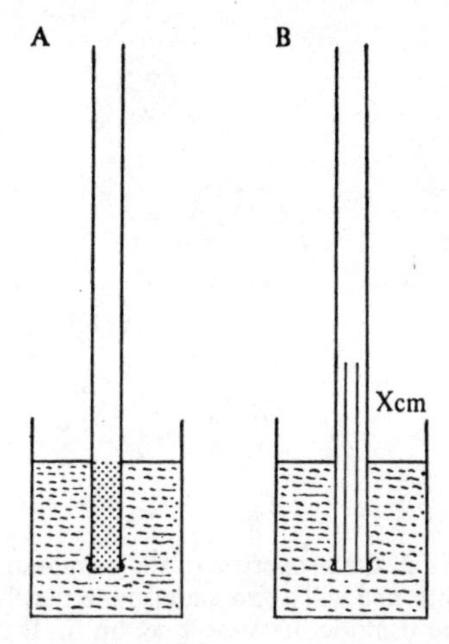

Fig.7.2. Osmotic pressure experiment (see text). The fluid in the beakers is a protein solution containing 5 g/l. In cylinder A is a protein solution containing 5 g/l while in cylinder B the solution contains 10 g of protein per litre.

What would happen if we repeated the experiment but instead of water in the beakers we put a protein solution containing 5 g of protein per litre? In situation A the protein solution in the cylinder and in the beaker would be identical. In situation B the protein solution in the cylinder would be twice as concentrated as that in the beaker. Again suppose that we leave the experiment set up for 24 hours. This time in cylinder A the fluid level will not have risen at all: no osmotic pressure is therefore set up when a membrane separates two solutions which are equal in concentration. In cylinder B the fluid level will have risen to half of what it was before. The osmotic pressure set up therefore depends on the concentration difference between two solutions on either side of a semi-permeable membrane. When the concentrations are equal, no osmotic pressure develops. When they are different, the bigger the concentration difference the higher is the osmotic pressure.

Properties of the membrane

Two further experiments can demonstrate that the membrane is as important as the solutions it separates in determining the osmotic pressure. Suppose that we repeat the first experiment but that this

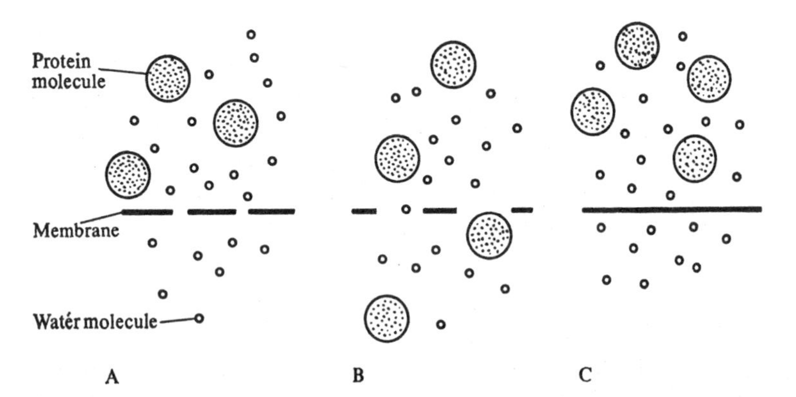

Fig.7.3. The importance of the properties of the membrane in setting up osmotic pressure. In A the membrane will allow water molecules but not protein ones to pass through and so an osmotic pressure is set up. In B the membrane is freely permeable to both water and protein and in C it is permeable to neither. In neither B nor C is an osmotic pressure set up.

time we use a membrane which allows protein molecules as well as water to flow freely across it. The protein molecules will pass from the cylinder into the water in the beaker and eventually the protein concentrations in the cylinder and the beaker will become equal because of a process of diffusion. No osmotic pressure will be set up.

Secondly, suppose we again repeat the experiment but this time use a rubber membrane which is completely impermeable to both protein and water. Again no osmotic pressure will be established. This time, of course, no protein will move into the beaker and the fluids in the cylinder and beaker will not change in composition.

The properties of the dissolved solids

Protein solutions were chosen to illustrate the concept of osmotic pressure because plasma is a protein solution and as we shall see later the osmotic pressure of plasma is very important in medicine.

But in fact any dissolved solid can exert an osmotic pressure provided that it cannot cross a membrane. Thus glucose, fructose, salt, sodium bicarbonate and all other materials which will dissolve in water can exert an osmotic pressure dragging one water from one side of a semi-permeable membrane to another. The osmotic pressures exerted by two different solutes dissolved in the same solution are additive. Thus a solution containing 1 gram-molecule of glucose per litre will exert the same osmotic pressure as one containing 1 gram-molecule of fructose per litre. One containing 2 gram-molecules per litre of glucose will exert the same osmotic pressure as one containing 1 gram-molecule of glucose and 1 gram-molecule of fructose per litre. What matters in the development of osmotic pressure is the *number* of dissolved particles: their chemical nature does not matter so long as they cannot cross the membrane concerned.

Summary of osmotic pressure

At this point it may be helpful to summarize concisely the conditions which must be fulfilled before an osmotic pressure can be established.

1. The membrane concerned must allow water to pass through freely but must not permit the passage of some substance dissolved in that water. No osmotic pressure can be set up if the membrane is freely permeable both to the water and to the solute nor if it is completely impermeable to both.

2. There must be a concentration difference between the two solutions on the two sides of the membrane. If such a concentration difference exists, water passes across the membrane from the weaker solution to the stronger one, thus tending to make both equal.

3. The size of the osmotic force set up depends on the magnitude of the concentration difference between the two solutions. The bigger the concentration difference the bigger will be the osmotic force.

4. The osmotic forces depend on the concentrations of dissolved particles per litre and not on the chemical nature of those particles. A solution of sodium chloride on one side of a membrane can be osmotically balanced by a solution of glucose, or potassium bicarbonate or protein or any other solute on the other side of the membrane provided only that the two solutions contain the same number of dissolved particles per litre.

Osmotic pressure and capillary function

Osmotic pressure is no mere academic concept. It plays a vital role in the functioning of the cardiovascular system. The capillaries are the tiny blood vessels with walls one cell thick where all the interchange between the blood and the extracellular fluid takes place. The capillary walls are permeable to water and almost all the substances dissolved in the plasma. The inorganic substances like sodium, chloride, potassium and bicarbonate can pass completely freely from blood to the extracellular fluid and most organic substances like glucose can pass freely in the same way. The only exceptions to this rule of free and easy movement between the extracellular fluid and the plasma are the molecules of the plasma proteins. These are much larger than any of the other molecules dissolved in the blood and they cannot escape into the extracellular fluid.

If the circulatory system is to function properly, the blood volume must be kept approximately constant. It would obviously be disastrous if all the plasma escaped from the blood vessels into the tissues. It would be equally disastrous if all the fluid in the tissues moved into the blood vessels and hopelessly overloaded the system. In some way the two fluid compartments must be kept in balance even though water and most of the substances dissolved in it can pass freely from one compartment to another.

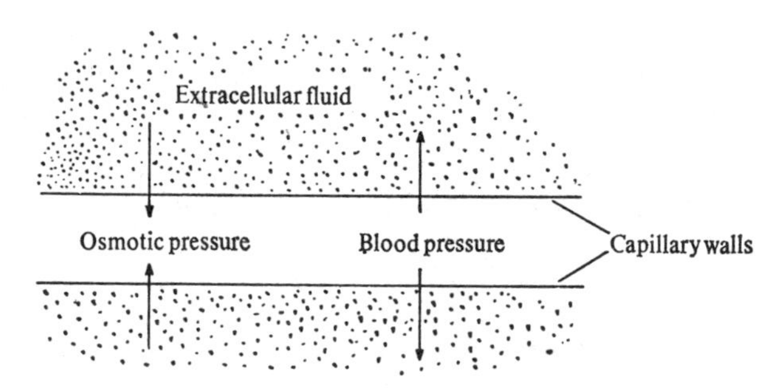

Fig.7.4. In a normal capillary there is an approximately even balance between the blood pressure pushing fluid out and the osmotic pressure of the plasma protein drawing fluid in.

If the blood is to be pushed around the body by the heart, the pumping mechanism must raise the pressure of the blood (see chapter 10). The pressure of the blood in the vessels is therefore greater than the pressure of the extracellular fluid outside the vessels. The situation is comparable to that of a rubber pipe through which ink is being pumped passing through a tank of clear water. The pressure of the ink inside the pipe is much greater than the pressure of the water outside. If a leak developed in the rubber pipe the ink would spurt out into the clear water and discolour it. Since the pressure of the blood is much higher than the pressure of the extracellular fluid and since the capillaries are extremely 'leaky' as far as the plasma is concerned, why is it that all the plasma does not escape?

The reason of course is that the blood pressure pushing fluid out is opposed by the osmotic pressure of the plasma proteins which tends to pull fluid into the capillaries. The two forces just about balance out so that normally there are no large shifts of fluid from blood to extracellular fluid or vice versa.

Osmotic pressure in some disease situations

The importance of osmotic pressure and capillary function is well illustrated by four clinical conditions.

1. EXTENSIVE BURNS. One of the most important effects of a burn is that capillary walls are damaged so that they are no longer impermeable to plasma proteins. All the constituents of plasma can therefore pass freely into the extracellular fluid and there is no osmotic pressure drawing fluid back into the capillaries. As a result there is nothing opposing the blood pressure and so large amounts of plasma are lost from the damaged capillaries. The blood volume becomes so low that the cardiovascular system is in danger of collapsing. This is why large transfusions of plasma may be life-saving in severe burns.

2. HAEMORRHAGE. When a large amount of blood has been lost, the heart can no longer maintain a normal blood pressure in the capillaries. The pressure of the blood in capillaries therefore falls below the plasma protein osmotic pressure. As a result fluid tends to move into the blood vessels so helping to restore blood volume. This is an important emergency method of helping the body to survive the crisis.

3. STARVATION. During prolonged starvation the concentration of the plasma proteins falls. The osmotic pressure they exert therefore

falls as well. The blood pressure may therefore be greater than the osmotic pressure and fluid moves out from the blood vessels into the tissues which become puffy and may give a false appearance of fatness.

4. HEART FAILURE. If the heart cannot pump all the blood it receives, that blood becomes dammed up in the veins and capillaries. The veins and capillaries become swollen and the pressure in them tends to rise. It exceeds the plasma protein osmotic pressure and therefore fluid again moves out into the tissues giving the puffy state known as oedema.

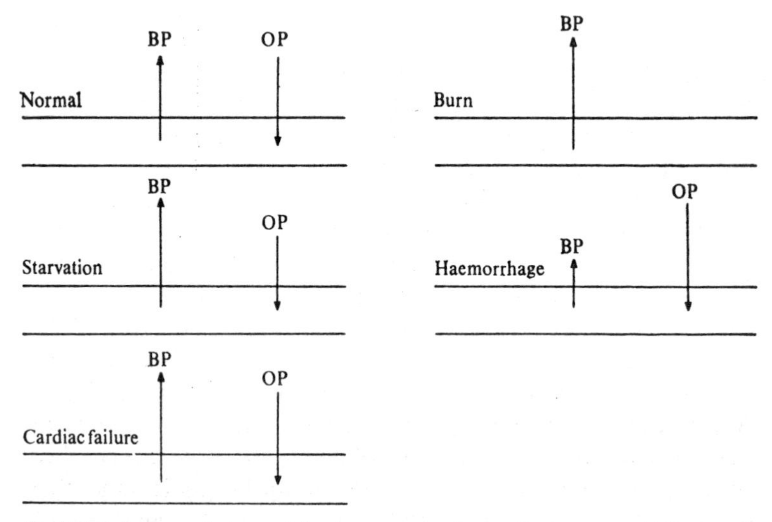

Fig.7.5. Imbalances between blood pressure and osmotic pressure which occur in various disease states (see text).

Osmotic pressure and cells

The cell membrane is also a semi-permeable membrane but it is a much more complex one than is the capillary wall. The cell membrane is much less permeable to many substances both organic and inorganic. Sodium ions, for example, cannot enter cells freely while potassium ions cannot leave and so they exert an effective osmotic pressure across the cell membrane. However, although the two solutions, the intracellular and extracellular fluids, are so different in composition, the total osmotic pressure which each one exerts across the cell membrane is equal to the total osmotic pressure exerted by the other. Because of this balance there is no overall tendency for fluid to move into or out of cells and the amounts of water inside the cells and in the extracellular fluid remain approximately constant.

If the osmotic pressure outside a cell is reduced—this can be most easily done artificially by immersing red cells from the blood in dilute salt solution instead of plasma—water will move from the dilute solution into the cells. The cells will swell and eventually burst. This occurs because the osmotic pressure exerted by the fluid inside the cells is greater than that exerted by the dilute solution outside. Water therefore must move into the cells. In the cases of red cells this process is known as haemolysis and all the red pigment haemoglobin escapes from the cells. In contrast, if red cells are immersed in a solution which has a greater osmotic pressure than normal, water is drawn out of the cells which shrink and become crinkled.

It is therefore important that red cells which are being kept for transfusion or for laboratory examination should always be kept in fluids which have the same osmotic pressure as normal plasma. When two fluids exert the same osmotic pressure they are said to be isotonic to one another. Isotonic saline is therefore a salt solution which exerts the same osmotic pressure as normal plasma. In medical circles isotonic saline is sometimes called 'normal' saline. This is chemically incorrect as normal saline in chemistry means a solution which contains 1 gram-molecule of salt dissolved in 1 litre of water. This is very much more concentrated than isotonic saline whose concentration is just under 1 per cent of that of normal saline on the chemical definition. It is therefore better to use the word isotonic rather than the word normal for a salt solution which has the same osmotic pressure as plasma.

8

Acids, bases and buffers

Most nurses shudder if they are asked a question about acids, bases or buffers and one can hardly blame them. Much of the complex terminology was devised many years ago and is now out of date. This chapter attempts to correct the situation by giving a simple modern account of the subject.

What is acidity?

The acidity of a solution is determined by the concentration in it of free hydrogen ions (H^+). That may seem a very simple definition but it is all that a nurse or doctor needs to know and if you keep it in mind it will greatly simplify your understanding of acids, bases and buffers.

The acidity of a fluid is extremely important in biology and medicine because the concentration of hydrogen ions markedly affects the behaviour of many of the chemical reactions which occur in the body. Increasing or decreasing the hydrogen-ion concentration away from its usual level causes these reactions to work abnormally rapidly or abnormally slowly. If the changes in acidity are extreme, proteins may be permanently damaged and cells may be destroyed. It is therefore vitally important that changes in the hydrogen-ion concentration of the body fluids should be kept to an absolute minimum.

For convenience, scientists have devised a way of measuring the hydrogen-ion concentration which is known as the pH scale. The centre point on this scale is pH 7 which represents the hydrogen-ion concentration of plain water. This pH of 7 is said to be neutral. When the hydrogen concentration increases above that in plain water, the solution is said to be acid. Paradoxically, its pH is said to fall. There are good chemical reasons for devising the pH scale in this paradoxical way but you do not need to understand them. All you need to remember is that when the pH falls this means an increase in hydrogen-ion concentration and an increase in acidity.

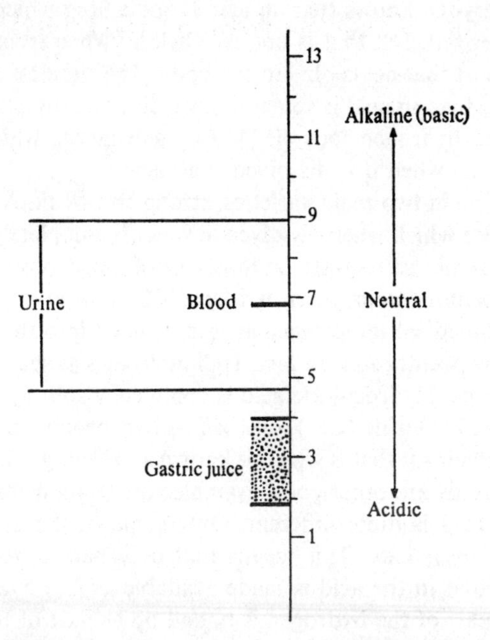

Fig.8.1. The pH scale showing the pHs of some important body fluids.

In contrast, when the hydrogen-ion concentration is less than that of pure water, the solution is said to be alkaline and its pH is above 7. The higher the pH the greater is the alkalinity and the lower is the hydrogen-ion concentration.

It is useful to have a rough idea of the pH of some important body fluids. The blood and the cerebrospinal fluid which bathes the brain have the most nearly constant pH. It is usually 7·4 and rarely fluctuates outside the limits of 7·3 to 7·5. The gastric juice, which is highly acid, has a pH of 2–3. The urine pH varies with the needs of the body. If the blood is too alkaline, alkali must be excreted and the pH may be 8 or even higher. If the blood is too acid, acid must be excreted and the pH may fall as low as 4·5. In general, meats contain much more acid than vegetables and so vegetarians tend to have a neutral or even alkaline urine while in meat-eaters the urine pH tends usually to be around 5–6.

What is an acid?

Almost everyone knows that an acid is something which turns blue litmus paper red. But that is not very helpful when trying to understand the part that acids play in the body. The simplest and clearest definition is that an acid is something which when dissolved in water releases free hydrogen ions (H^+). Any substances which provides hydrogen ions when it is dissolved is an acid.

Acids come in two main varieties, strong and weak. A strong acid is a substance which when dissolved in water is completely broken up (dissociated) into its constituent ions, one of those ions being hydrogen. Hydrochloric acid, often written HCl, is an excellent example which is of medical importance as it is secreted into the stomach in quite high concentration. In fact, HCl molecules as such do not exist in gastric juice. Hydrochloric acid is completely split up into hydrogen (H^+) and chloride (Cl^-) ions. *All* its hydrogen is in the form of free hydrogen ions: that is why it is known as a strong acid. Sulphuric and nitric acids are other good examples of strong acids.

A weak acid is quite different. Only some of the molecules are split up to form ions. This means that only part of the hydrogen which is found in the acid is made available as free hydrogen ions. The other part of the hydrogen is bound up as part of the molecule and so does not affect the acidity of the solution. The behaviour of any weak acid can be accounted for by considering the hypothetical weak acid, HA

$$HA \rightleftharpoons H^+ + A^-$$

Some is in the form of HA molecules in which the hydrogen is bound to A and is not free in ionic form: it therefore contributes nothing to the acidity of the solution. On the other hand some of the HA breaks up to give H^+ and A^- and these free hydrogen ions can affect the acidity of the solution. Weak acids are therefore called weak because only part of their hydrogen is available in the form of free hydrogen ions. Good and medically important examples of weak acids are carbonic acid, haemoglobin, phosphoric acid and acetic acid.

What is a base (alkali)?

Alkali and base are two terms which effectively mean the same thing. As we have seen it is the concentration of free hydrogen ions which determines the acidity or the pH of a solution. An acid is a substance which can add hydrogen ions to a solution and a base is the opposite of an acid. Bases are therefore substances, often negatively charged ions, which can react with hydrogen ions and so remove them from solution and inactivate them. Biologically speaking, the most important bases are bicarbonate, phosphate and haemoglobin ions, the ammonia molecule and the carboxylate and amino groups attached to protein molecules. The ways in which they work are shown in figure 8.2.

Buffers

If you study the reactions shown in figure 8.2 you will notice something very interesting. All the substances on the right-hand sides of the chemical equations can react with hydrogen ions so that the

H_2CO_3 Carbonic acid	$\leftrightharpoons$	H^+	+	HCO_3^- Bicarbonate
HbH Haemoglobin	$\leftrightharpoons$	H^+	+	Hb^- Haemoglobin ion
$H_2PO_4^-$ Dihydrogen phosphate	$\leftrightharpoons$	H^+	+	$HPO_4^=$ Monohydrogen phosphate
$-COOH$ Carboxylic acid group	$\leftrightharpoons$	H^+	+	$-COO^-$ Carboxylate ion
$-NH_3^+$	$\leftrightharpoons$	H^+	+	$-NH_2$ Amino group
NH_4^+ Ammonium ion	$\leftrightharpoons$	H^+	+	NH_3 Ammonia
WEAK ACIDS				BASES

Fig.8.2. The important weak acids and bases in the body. Together these act as buffers (see text).

hydrogen no longer exists in free ionic form (H^+). All these substances on the right-hand sides of the equations are therefore bases. The substances on the left-hand side of the equations are, however, quite different: they are all weak acids. When they dissociate they liberate free hydrogen ions and therefore make a solution more acid. In the process they also form bases. Therefore in summary, a weak acid is a substance which can split up to give hydrogen ions and a base. A base is a substance which can react with hydrogen ions to give a weak acid.

Substances which can interact with hydrogen ions in this way are known as buffers and they are very important in medicine. They are called buffers because they have the property of resisting or of 'buffering' changes in the hydrogen-ion concentration of a solution. This means that when we add hydrogen ions to a solution containing a buffer or remove hydrogen ions from such a solution, the change in pH is always less than we might otherwise expect if no buffer were present.

Any buffer consists of an acid, a base and hydrogen ions. This system acts to resist changes in pH in the following ways:

1. When hydrogen ions are added to such a system, not all the hydrogen ions which are added remain free and able to change the pH of the solution. Some of the hydrogen ions which are added combine with the base and are effectively inactivated. The solution does become more acid but the change is not as great as might have been expected. For example if we add one-hundred hydrogen ions to a solution which contains no buffer system, all the hydrogen ions will be effective in increasing the acidity of that solution. But if we add one-hundred hydrogen ions to a solution containing a buffer, seventy might be inactivated by combination with the base leaving only thirty free to alter the solution's pH.

2. When hydrogen ions are removed from a buffer system, some of them are replaced by the splitting up of some of the molecules of the weak acid part of the buffer. If one-hundred hydrogen ions are removed, seventy might be replaced by the splitting up of weak acid molecules to give free hydrogen ions. Again the drop in hydrogen-ion concentration is much less than would be predicted from knowing only the number of hydrogen ions removed. Again the change in pH is less than would have occurred had no buffer system been present.

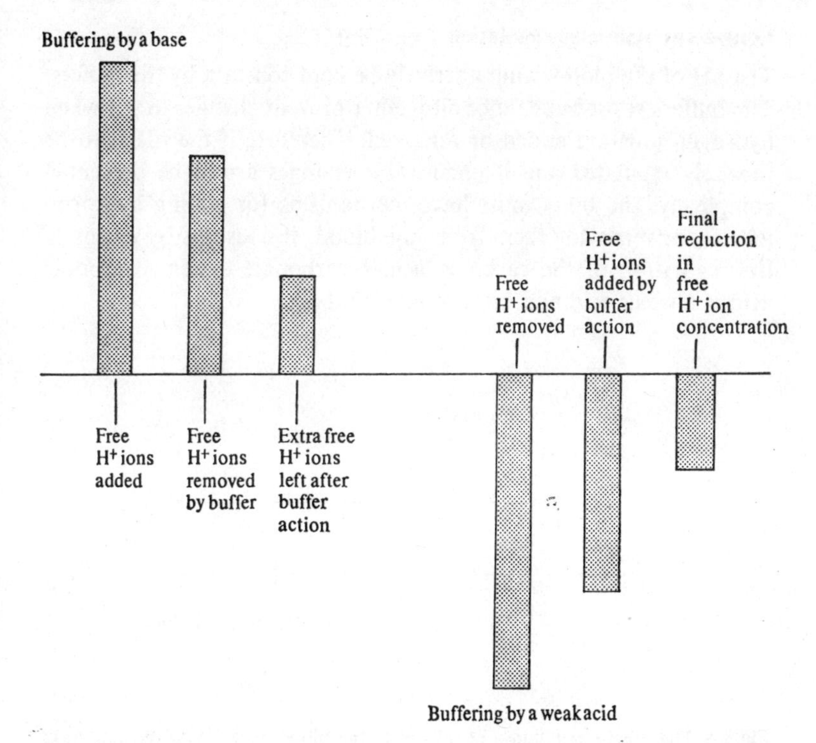

Fig.8.3. The actions of buffers which diminish the expected pH changes when hydrogen ions are added to a solution or removed from it (see text).

Importance of buffers

In the first section of this book we saw the need for the maintenance of the constancy of the pH of the blood. During life this constancy is perpetually being assaulted. Acids and bases are added to the body in the food and removed by the kidney. Acids, and in particular carbon dioxide, are added to the blood by all active organs. It is therefore clear that there are excellent reasons why the blood should be rich in buffers in order to reduce the impact of these changes. In blood the most important buffers are bicarbonate, haemoglobin and the carboxylate and amino groups of the plasma proteins. In urine the permissible range of pH is much wider but even there the urine must not become too acid. The main buffers in urine are bicarbonate, phosphate and ammonia. All these buffers perform the same function of helping to reduce the consequence of the addition of hydrogen ions to a solution or their removal from a solution.

Long-term acid-base regulation

The pH of the blood cannot actively be kept constant by the buffers. The buffers resist but cannot altogether prevent changes in pH when hydrogen ions are added or removed. Therefore if the pH is to be precisely regulated and if undesirable changes are to be prevented completely, the body must have mechanisms for adding hydrogen ions to, or removing them from, the blood. It does this by means of the regulation of the carbonic acid–bicarbonate system. Carbonic acid is a weak acid and bicarbonate is a base.

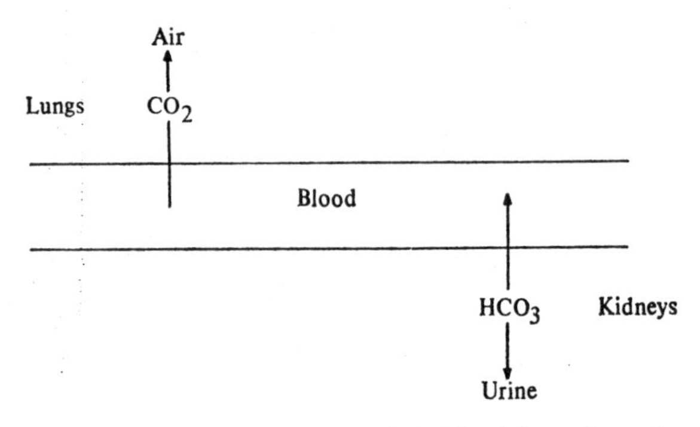

Fig.8.4. The precise regulation of the pH of the blood depends on the actions of the lungs which determine the concentration of the acid carbon dioxide and the kidneys which determine the concentration of the base bicarbonate.

The lungs determine the carbonic acid concentration of the blood. This is because carbonic acid is manufactured by the reaction with water of the gas carbon dioxide. Carbon dioxide is produced by the cells in large amounts and is removed from the blood by the lungs. If breathing becomes slow, carbon dioxide accumulates and the carbonic acid concentration of the blood rises: some of this carbonic acid splits to release hydrogen ions and the blood therefore becomes more acid. If breathing is rapid, carbon dioxide is removed from the blood more quickly and its concentration falls. The concentrations of carbonic acid and hydrogen ions also fall in consequence and the blood becomes more alkaline. Thus by precisely regulating the rate of breathing the concentration of carbon dioxide and therefore of carbonic acid in the blood can be kept constant.

The other important organ which regulates hydrogen-ion concentration is the kidney. If the blood is too acid, the kidney manufactures bicarbonate ions which are then added to the blood: because bicarbonate is a base it combines with hydrogen ions and inactivates them, so reducing the acidity of the blood. If the blood is too alkaline the kidneys in contrast remove bicarbonate ions and excrete them in the urine. Thus between them the kidneys and the lungs keep the pH of the blood remarkably constant.

Section 3

PHYSICS

9

Heat

The aim of this section is not to teach physics to nurses but to describe those very few aspects of physics which are genuinely relevant to medicine. Let us start by considering heat.

Heat is a commodity which it is easiest to understand by considering the effects which it has. The most important actions of heat which can affect biological systems are as follows:

1. It raises the temperature of objects into which it flows. When heat flows from an object the temperature of that object falls.

2. When the temperature of chemical systems rises, chemical reactions take place more rapidly than usual. Roughly speaking when the temperature of a cell rises by 10°C, the reactions occurring within it occur at double the rate. In contrast, when cells are cooled, the reactions within them are slowed down.

3. Most substances expand when their temperature rises. That is to say a given weight of a hot object occupies more space than the same weight of that object when it is cold. In general, liquids expand more than solids and gases expand more than liquids.

4. If the temperature of proteins in the body rises too high, the proteins will be destroyed. The cells cannot then survive and so the whole body will die.

The movement of heat

Heat moves around the body and to and from the body in four main ways.

1. CONDUCTION. This occurs when two objects are actually touching. Heat always flows from the hotter object to the colder one. When you get into a cold bath, for example, heat flows by conduction from your body to the bath water. When you get into a very hot bath, heat

flows by conduction from the hot water into your body. Two factors determine the speed with which heat is transferred from your body to the water or vice versa.

a. The difference in temperature between your body and the water. If your body is the same temperature as the water, no

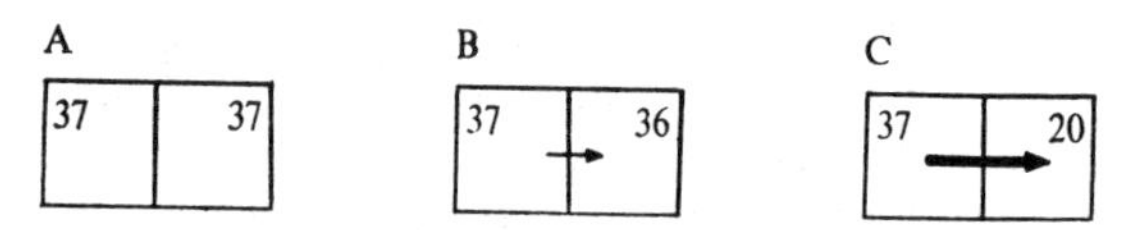

Fig.9.1. The rate of heat transfer between two objects depends partly on the size of the temperature difference between them. The figures are in degrees centigrade.

heat will flow from one to the other. If your body is 1° warmer than the water, heat will flow very slowly from your body to the water. If your body is 10° warmer, heat will flow away ten times more rapidly and you will feel that the water is much colder. The bigger the temperature difference, therefore, the more rapid is the heat flow.

b. The amount of surface contact between the two objects. If you put just one hand in a bath of cold water, much less heat will flow from your body to the bath than if your whole body is immersed. The amount of heat transferred therefore also depends on the amount of surface contact between the two objects.

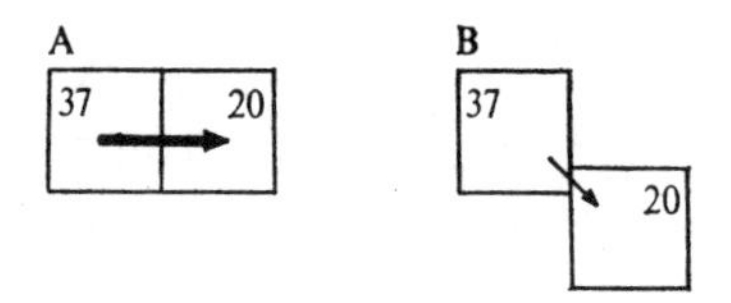

Fig.9.2. The rate of transfer also depends on the amount of surface contact between the two objects.

2. CONVECTION. When heat is conducted from your body to the air which surrounds it on a cool day, the temperature of that air rises. The air therefore expands and a given weight of hot air occupies much more space than the same weight of cold air. Looking at it another way, a given volume of hot air is considerably lighter than the same volume of cold air. The hot air therefore tends to float

upwards and as it rises it is replaced by cold air moving in from the sides. Convection helps to explain why in a cinema without air-conditioning the upper circle is always much hotter than the stalls.

3. TRANSPORT OF HEAT BY THE MOVEMENT OF A HOT OBJECT. This is biologically important in two main ways. The cells which are actively working tend to be hotter than the blood and so heat flows from the cells to the blood. It is then carried around the body by the movement of the blood. Similarly on a cold windy day, the warm air next to your body is blown away and replaced by cool air almost before it can be warmed.

4. RADIATION. This occurs when there is no physical contact between two objects. The best example is the warming of the earth by the heat rays from the sun travelling across 93 million miles of empty space. Any hot object, such as a fire, radiates heat to any cooler object in the vicinity. The rate of heat transfer by radiation depends on the same two factors which govern the rate of heat transfer by conduction. The greater is the temperature difference between two objects, the faster will heat radiate from one to the other. Similarly, the greater the surface area exposed, the greater will be the amount of heat transferred. You warm up much more rapidly when you put your whole body in front of a radiator than you do when you put a single hand in front of it.

Units of heat

In order to be able to understand heat fully it is essential to know how the quantity of heat can be measured. The unit usually used is the calorie. The calorie is the amount of heat which will raise 1 cm^3 of water through 1 °C. The calorie is a very small unit as 100 cal are needed to raise a single cubic centimetre of water from freezing point to boiling point. Because the calorie is so small, the kilocalorie is often used instead. One kcal is equal to 1,000 cal. The kilocalorie is sometimes referred to as a Calorie with a capital C but this produces so much confusion between calories and Calories that it is better to use the word kilocalorie and then there can be no doubt about what you mean.

The amount of food required daily by the body is often expressed in terms of the amount of heat which that food can release when it is oxidized in the body. The Calorie used in most diet sheets is in fact the kilocalorie. Therefore when a diet says that someone needs 2,000 Cal/day, this means that he needs 2,000 kcal. The term 'needs 2,000 kcal' is really shorthand for saying 'needs that amount of food which when oxidized in the body is capable of producing 2,000 kcal of heat'.

Heat production

Many of the chemical reactions which occur in the body give off heat. This is particularly true of the chemical reactions which occur during muscular activity. The heat produced in these reactions helps to maintain the temperature of the blood at 37 °C (about 98 °F) which in most parts of the world is well above the temperature of the surrounding environment.

In cold weather the body often speeds up its chemical activity. The extra heat is required to counteract the increased rate of heat loss to the cold environment. You yourself are well aware of this although you may not have thought about it before. When you feel cold you often consciously decide to walk more quickly or to do some other exercise in order to warm up. If this does not increase your heat production sufficiently, your temperature-regulating centre in the brain may take over and order your muscles to start working involuntarily: in other words you start shivering and the extra heat produced helps to warm you up.

Normally all the heat in your body comes from your own chemical activity. However, there are some situations in which heat enters the body because the environment is hotter than the body itself. The common place where this happens in most people's experience is the bath. Workers in some types of factories and in ships' engine rooms may experience these conditions for long parts of the day. In some regions of the world the climate may be such that the outside air temperature is hotter than that of the body, but the hot period usually covers only part of the day. In every part of the world, for the great majority of the time, the heat in the body comes not from outside but from the body's own working.

Heat loss

Heat is lost from the body in several different ways. At the beginning of this discussion it is essential to realize that heat can be lost from the body only by crossing the surface of the body. Therefore before it can leave the body the heat must be transported from all the deep structures where it is produced, to the skin. It is transported to the skin by the blood.

Once heat reaches the skin it can leave it by the mechanisms which we have already discussed.

1. By conduction of the heat to cooler objects and cooler air in contact with the body.

2. By convection. The air heated by contact with the body expands, becomes lighter and rises upwards. Cool air comes in to take its place.

3. By radiation of the heat to cooler objects in the vicinity which need not be in actual contact with the body.

The rate of heat lost by these three methods depends on the temperature difference between the body and its surroundings and the amount of the body surface exposed. Clothing which restricts the passage of heat can greatly reduce heat loss. The greater the temperature difference between the body and its environment, the more rapid will be the transfer of heat. This explains why you feel cold on a cold day. There is such a large temperature difference between your body and its surroundings that the heat flows away very quickly.

Conduction, convection and radiation are highly effective ways of losing heat, provided that the environment is cooler than the body. But as the temperature of the environment rises and the difference between it and body temperature decreases, the rate of heat loss from the body becomes less and less. When the environmental temperature rises above that of the body, heat is actually gained from the surroundings by the action of conduction and radiation. Under these conditions there is clearly a danger that the body temperature will become dangerously high, especially if the body's own production of heat is pushed up to high levels by exercise. In fact survival under these circumstances would be impossible unless the body had an alternative way of losing heat which did not depend on conduction, convection or radiation.

Fortunately there is such an extra way of losing heat. This is the familiar phenomenon known as sweating. Sweating depends on the peculiar fact that when 1 cm^3 (1 ml) of water evaporates it takes from its surroundings an enormous quantity of heat. This is known as latent heat and it amounts to about 540 cal for every cubic centimetre of water which evaporates. This means that when 1 cm^3 of sweat evaporates, over 500 cal of heat are removed from the underlying skin and the blood which supplies that skin. This is true even if the surrounding air is hotter than the body. This means that even when

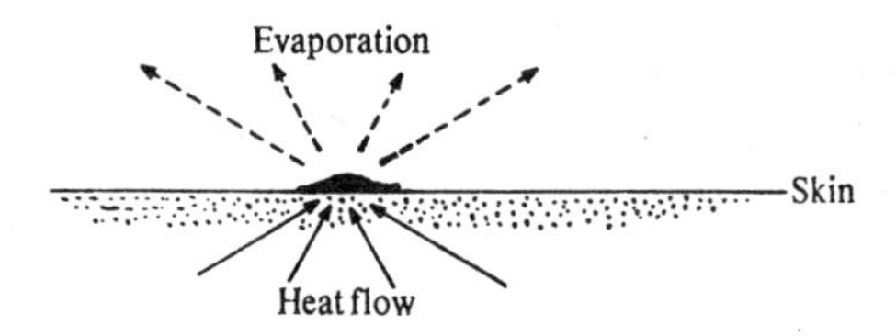

Fig.9.3. When water evaporates from the surface of the body it draws heat from the underlying skin and the blood flowing through it.

all the other methods of losing heat are ineffective, the body can still lose heat by sweating. The water is secreted onto the surface of the skin by special glands known as sweat glands and as it evaporates the body is effectively cooled.

The rate at which water evaporates can be increased by raising its temperature (as everyone knows who has by accident boiled a kettle dry) but this is not very important in raising the evaporation rate during sweating as the temperature of the sweat is always very close to the temperature of the body. Two other factors are however important in altering the rate of evaporation and thus the effectiveness of sweating as a heat-loss mechanism.

1. AIR MOVEMENT. When air moves over a water surface, as on a windy day, the rate of evaporation is greatly increased. Every housewife likes to hang out her washing on a windy day because then it dries so much more quickly. This means that on a windy day the rate of heat loss from water evaporating from the skin is greatly increased. This accounts for the fact that hot, still days are much less comfortable than hot, windy days. On a hot windy day the sweat evaporates very rapidly and cools the body quickly and effectively. It also explains why you feel so much colder on coming out of the water after swimming on a windy day than you do on a still day. On a windy day, the water on your skin evaporates very rapidly and cools you down quickly.

2. HUMIDITY. Just as solids can apparently disappear without trace when they dissolve in water, so water molecules can 'dissolve' in air.

Just as there is also a limit to the amount of solid which water can carry in solution, so there is a limit to the amount of water which the air can carry before the excess is precipitated out as rain or condensation. When air is carrying the maximum amount of water vapour

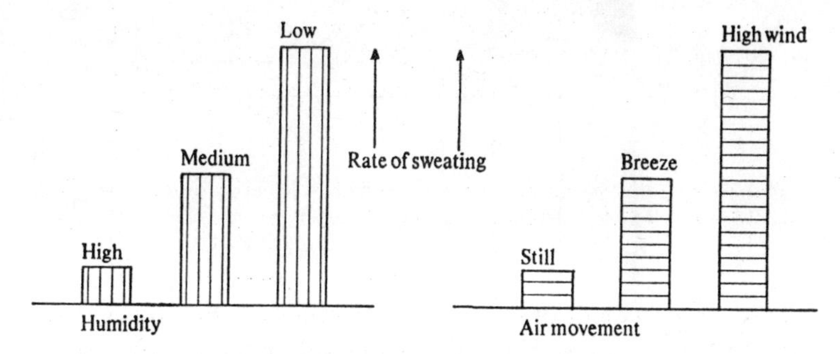

Fig.9.4. Factors which affect the rate of sweating (see text).

which it can hold, it is said to be 100 per cent saturated with water vapour: the relative humidity of the air is said to be 100 per cent. Under these conditions the air cannot carry any more water and so no water can evaporate. In contrast, when the air is very dry and carries no water at all, the relative humidity is said to be 0 per cent. It is obvious that evaporation will be much more rapid and sweating will be much more effective when the relative humidity is low. On a dry hot day, the sweat evaporates as soon as it is formed and you feel reasonably cool even though the temperature of the environment may be very high. But when the humidity is high, the sweat cannot evaporate easily: instead it forms large drops and runs off your skin without evaporating and without cooling you down. Such conditions are often said to be 'sticky' and you feel very uncomfortable even though the air temperature may be lower than when the weather is hot and dry.

Volume and surface area

When you nurse new-born infants, especially if they are premature and very tiny, you will be told that one of the major problems is to keep such tiny scraps warm. That is why they are so often kept in specially heated incubators. Part of the explanation for this is quite simple and can easily be understood in terms of elementary physics.

It depends on the fact that heat is produced by the tissues through the whole volume of the body but is lost only across the body surface.

In order that you may understand this it is best to take a simple example. Consider two cubes, one of which has sides 1 cm long and

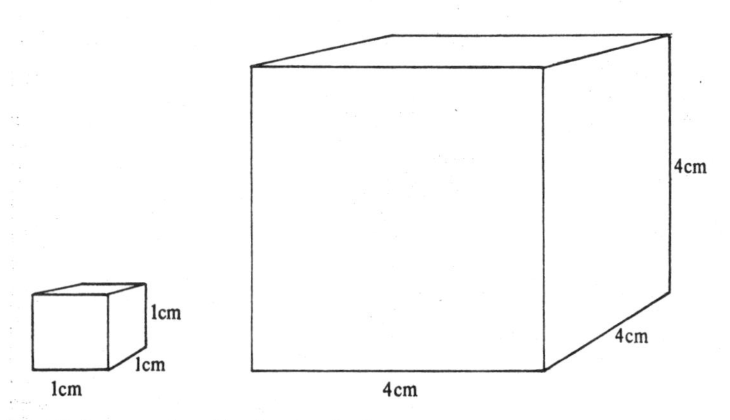

Fig.9.5. Large and small cubes (see text).

one of which has sides 4 cm long. The volume of each cube can be obtained by multiplying the length of each side by itself three times. So the volume of the large cube is $4 \times 4 \times 4 = 64$ cm^3 while the volume of the small cube is $1 \times 1 \times 1 = 1$ cm.^{3} The large cube therefore contains sixty-four times as much material as the small one: this is not far from the ratio of the mass of a large man to the mass of a small premature baby.

What about the areas of the surfaces of the two cubes? Each side in the small cube has a surface area of $1 \times 1 = 1$ cm^2. Each side of the large cube has a surface area of $4 \times 4 = 16$ cm^2. Both cubes have six sides and so the total area of the small cube is $1 \times 6 = 6$ cm^2 while the total area of the large cube is $16 \times 6 = 96$ cm^2.

What does this mean? It means that for every cubic centimetre in the large cube there is 1·5 cm^2 of surface. But for the cubic centimetre in the small cube there is 6 cm^2 of surface. Thus in relation to volume there is four times as much surface area in the small cube. If heat is produced by the volume of the tissue but lost across its surface, the cubic centimetre in the small cube will lose heat four times more rapidly than each cubic centimetre in the large cube. Small babies, like small cubes, have large surface areas in relation to their volume. They therefore tend to lose heat very much more rapidly than adults and soon their body temperature falls unless special precautions are taken to keep them warm.

10

Liquids

A knowledge of the basic principles involved in the pumping of liquids around the body could hardly be more important in medicine. The circulation of the blood depends entirely on such pumping and is obviously of central significance.

The pumping of the heart

If the action of any pump is to be understood, it is essential to know that liquids are incompressible. A given volume of water or of blood cannot be forced to occupy a smaller space simply by squeezing it. This can easily be demonstrated by drawing some water up into a syringe and sealing the end. If you then note how much water there is in the syringe and squeeze hard, you will find it quite impossible to squash the water into a smaller volume. And this is not just because your strength is limited. It is because by their very nature liquids cannot be compressed no matter how massive are the forces applied to them.

The action of all liquid pumps depends on this incompressibility. This is true of the heart just as it is true of man-made machines. In essence the heart is extremely simple. Each side of the heart contains one chamber—the atrium—whose function seems to be mainly the storage of blood ready for the next beat, and a second chamber—the ventricle—which acts as the real pump. Guarding the entry to, and exit from, each ventricle is a set of one-way valves. The valves which lie between the atrium and the ventricle will let blood pass in only one direction, from the atrium to the ventricle. In a normal person it is impossible for blood to flow back from the ventricle to the atrium. Immediately there is any tendency for the blood to flow this way, the valves slam shut.

The valves between the ventricle and the pulmonary artery on the right side, or the aorta on the left side, are also one-way. Blood can pass only out from the ventricle. It cannot normally return from the pulmonary artery or from the aorta into the ventricle.

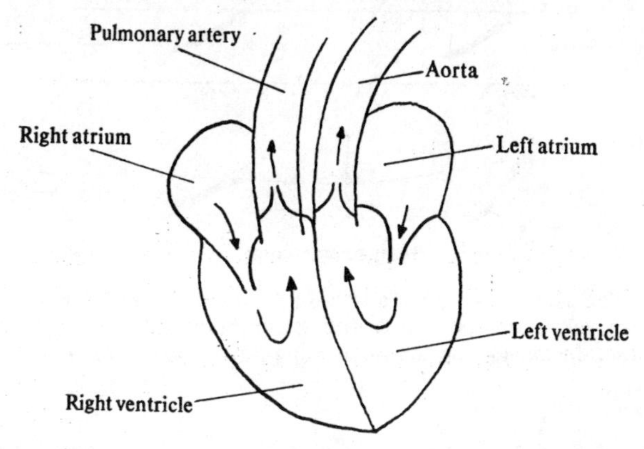

Fig.10.1. Outline sketch of the heart showing the vital non-return valves.

What happens therefore when the walls of the ventricle contract? They try to squash the blood, to make it occupy a smaller volume, but this they cannot do. Instead the pressure of the blood rises. The blood cannot go back from ventricle to atrium, it can only go onwards into the pulmonary artery or aorta. The valves therefore open and let the blood out. The pumping of the blood therefore depends on its incompressibility and the one-way action of the valves.

The venous pump

When the body is at rest, the heart is the only effective pump. The circulation of the blood depends upon it entirely. But during exercise other pumps come into action and help to ease the heavy load which would otherwise be placed upon the heart. These other pumps are in the thin-walled veins and they depend on the fact that most veins have numerous valves in them which allow blood to flow only towards the heart. Blood cannot normally therefore flow backwards in the veins. During exercise the veins are intermittently compressed by the actions of the working muscles. When a working muscle presses upon a vein, there is a tendency for blood to be squeezed from that section of the vein. The blood cannot go back. It can only go forwards and so the venous pumps help the return of blood to the heart.

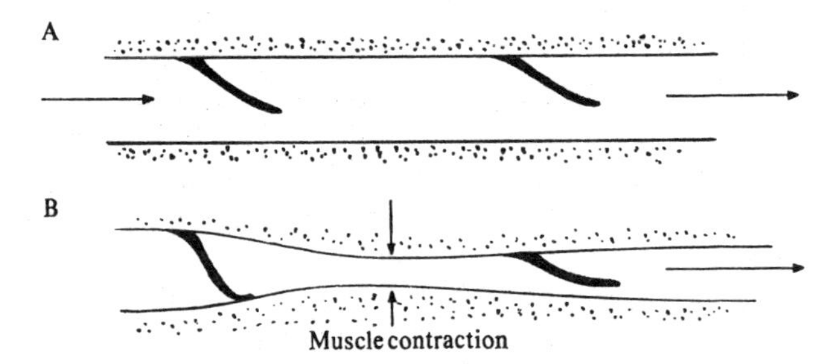

Fig.10.2. The pumping action of a vein. In A, at rest, the blood flows quietly on and all the valves are open. In B, during compression of the vein by muscular contraction, the valves stop the blood being pushed backwards and force it onwards.

The pressure produced by a pump

When a liquid is squeezed, as the blood in the ventricles is squeezed by the action of the heart muscle, the volume of the liquid cannot change. However, the squeezing does produce some change in the liquid, we say that it raises the pressure of the liquid. What do we mean by this?

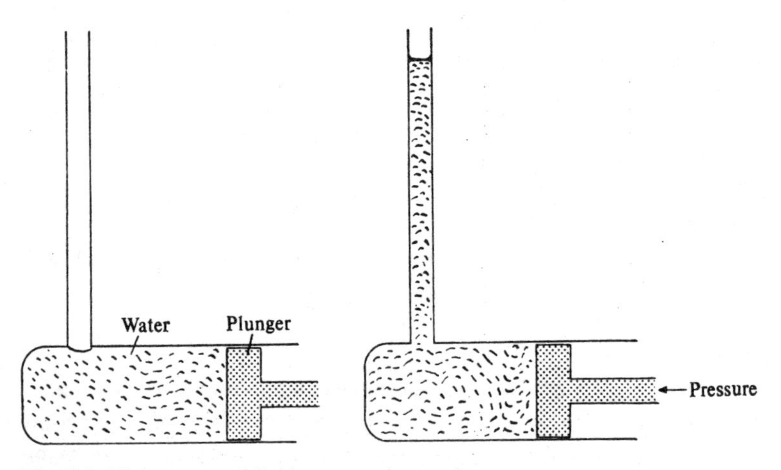

Fig.10.3. The concept of fluid pressure (see text).

Suppose we set up the experiment shown in figure 10.3. A barrel, fitted with a smoothly sliding plunger, is filled with water. Coming from the top of the barrel is a fine tube ascending vertically. What happens when the plunger is pressed? The water will be forced out of the barrel into the fine tube and the harder we press the higher the water will go. Suppose we press until the column of water in the tube is 136 cm (about 4 ft 4 in) high. We then say that the pressure of the water in the barrel is equal to 136 cm of water. By pressing on the plunger we have raised the pressure of the water in the barrel until it can hold up against gravity a column of water 136 cm high. If we then take the pressure off the plunger and if the plunger is smooth sliding, the weight of the column of water then pushes the water back into the barrel. This occurs until the height of the column of water is zero. We then say that the pressure of water in the barrel is zero.

Why did I choose the rather odd figure of 136 cm of water to illustrate this point. I did it because pressures are conventionally measured not in centimetres of water but in millimetres of mercury (Hg is the chemical symbol for mercury). Mercury is very much heavier than water, 13·6 times heavier in fact, and so a column of mercury is very much shorter than a column of water of equivalent weight. A column of water 136 cm high exerts a pressure equivalent to that exerted by a column of mercury 100 mm high. When design-

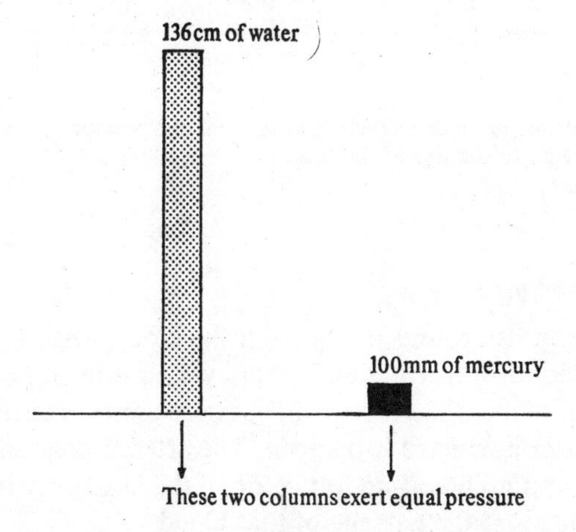

Fig.10.4. A column of water 136 cm high exerts a pressure equal to that exerted by a column of mercury 10 cm (100 mm) high. This is because mercury is 13·6 times heavier than water.

ing an instrument such as the sphygmomanometer which is used for the measurement of blood pressure it is much easier to make one with a short column of mercury than one which uses an inconveniently long column of water.

Although mercury, for convenience, is usually used when making pressure measurements, it is important to remember that blood is mainly water and that in order to pump blood up to the top of the head the heart must raise its pressure to at least 40 cm of water. In fact in a normal person the arterial pressure very rarely falls below about 60 mmHg which is equivalent to about 80 cm of water and so there is usually an ample margin of safety.

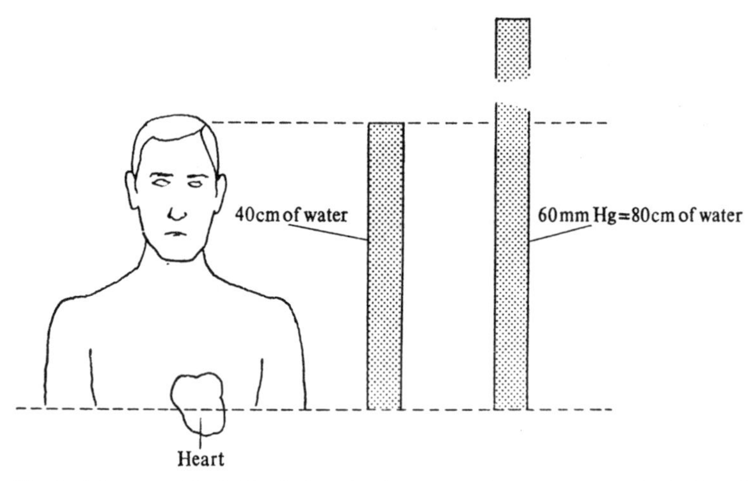

Fig.10.5. In normal individuals there is always enough pressure to push the blood from the heart up to the top of the head.

Resistance and pressure

The blood travels around the body in the blood vessels, a complex system of smooth-walled tubes. If these vessels offered no resistance to the flow of blood, the heart could not raise pressure to the required level no matter how hard it pumped. The arterial pressure is sufficiently high to pump blood up to the top of the head only because the blood vessels resist the passage of that blood.

You can demonstrate the effect of resistance to flow yourself if you have access to a tap (preferably a garden one!) which has a rubber pipe attached to it. Turn the tap half-way on. If the pipe is wide open and you hold it so that the end of it is facing upwards, the water will probably come out without any great force: its pressure will be low. But if you compress the pipe between your fingers, so

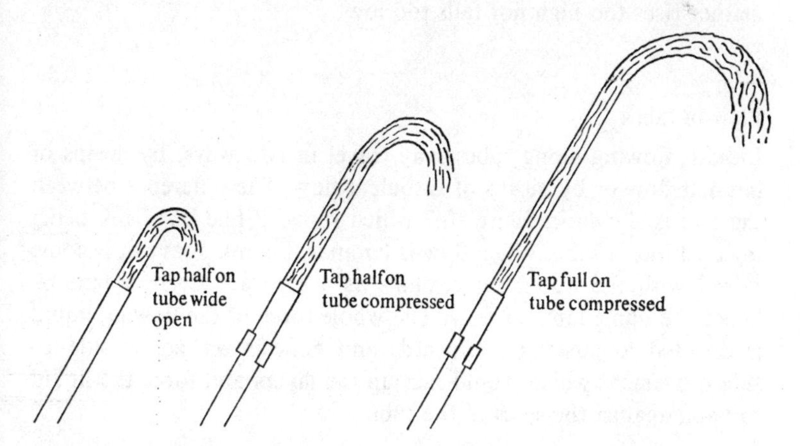

Fig.10.6. Factors controlling the pressure of liquid in a tube (see text).

greatly reducing the diameter of the nozzle through which the water must flow, you can convert the dribble into a powerful jet. By greatly increasing the resistance to flow, you have increased the pressure of the water. Now suppose that while still compressing the pipe you turn the tap full on. The water jet will then shoot even higher up into the sky: you can therefore increase the pressure by increasing the rate at which water flows into the tube. It is possible to alter pressure in a liquid by either altering the rate of flow or by altering the degree of resistance to that flow.

How does this compare with the situation in the body? There is a very close parallel. Many of the blood vessels are relatively constant in diameter and therefore the resistance which they offer to the flow of blood changes little. However, as the arteries branch into smaller and smaller vessels they give rise to the tiny arterioles which come just before the capillaries. The artioles have a thick muscular wall: both nerves and hormones can alter the diameters of arterioles by altering the degree of muscular contraction. Therefore the body, by opening or closing the arterioles, can readily alter the resistance to the flow of blood.

On the other hand, the heart is not an inflexible pump which pours out a fixed amount of blood every minute. It can readily alter its rate of pumping. Therefore by controlling the output of the heart (cardiac output) and the resistance offered by the arterioles (peripheral resistance), the body can adjust the arterial pressure so that it neither rises too high nor falls too low.

Flow in tubes

Liquids flowing along tubes may travel in two ways, by means of laminar flow or by means of turbulent flow. The difference between the two is shown in figure 10.7 which shows a fine jet of ink being injected into a tube. When flow is laminar, the ink does not become mixed with the water but remains as a separate layer or lamina, hence the name laminar flow. The whole force of the flowing liquid is directed to pushing it onwards and none is wasted on side-to-side movement which would mix up the layers and force the liquid to bang against the sides of the tube.

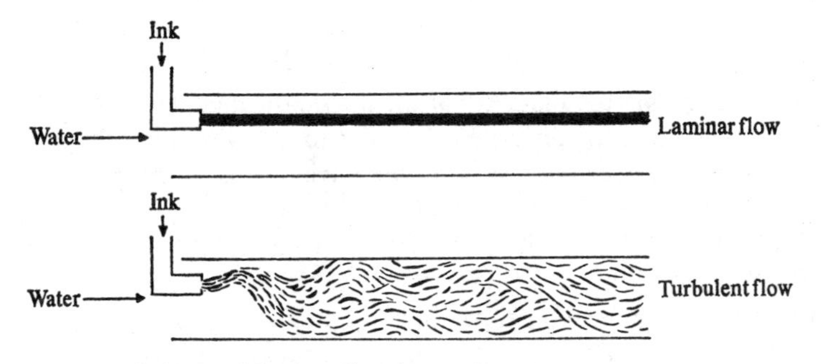

Fig.10.7. Turbulent and laminar flow (see text).

Turbulent flow is quite different and its nature can be guessed at from its name. Immediately on entering the tube the ink is thoroughly mixed up with the water. Much of the force of the flowing liquid is wasted by side-to-side movement of the water and in impacts upon the walls of the tube. Thus the liquid flowing along the tube rapidly becomes thoroughly mixed.

Three main factors determine whether flow will be laminar or turbulent.

1. The diameter of the vessel. Turbulent flow becomes much more likely as a tube becomes narrower.

2. The speed of flow. The likelihood of turbulence increases as the speed of flow increases.

3. The presence of irregularities in the vessel wall or of sudden changes in the direction of the tube.

This would all be highly academic as far as nursing is concerned but for one thing. This is that when you put a stethoscope over a vessel in which turbulent flow is occurring you can hear a rushing sound or 'murmur'. If you put a stethoscope over a vessel where laminar flow is occurring you can hear nothing. This is because the side-to-side movement of the fluid in turbulent flow sets the walls of the vessel vibrating. These vibrations are picked up by the stethoscope as sounds. A murmur therefore can occur whenever flow is turbulent. Blood flow in the cardiovascular system is normally laminar but turbulence commonly occurs in the following situations:

1. It is quite common in the aorta in young children because the blood is pumped rapidly out of the heart into a vessel which is relatively narrow. This type of turbulence indicates nothing abnormal. A similar type of normal turbulence may occur during and after severe exercise when the output of the heart is greatly increased and blood is ejected much more rapidly into the aorta than is normally the case at rest.

2. Flow past the wide open valves of the heart is normally laminar but if the valves are damaged or narrowed (stenosed), turbulence may occur giving a murmur which indicates some abnormality.

3. Flow in large arteries is normally laminar but if there is a large fat deposit as in atheroma, the vessel may become so narrowed and its surface so irregular that turbulence occurs. Murmurs cannot normally be heard over the carotid arteries in the neck or the renal arteries in the abdomen. Murmurs over these areas almost certainly indicate atheroma.

It is therefore clear that an understanding of the principles of the flow of liquids in tubes can considerably increase one's understanding of medicine.

Surface tension

Next time you blow a soap bubble for a small child, stop blowing before a large bubble has become detached from the pipe or ring which you are using. What will happen to the bubble? Invariably it will decrease in size until all the air has been emptied from it and

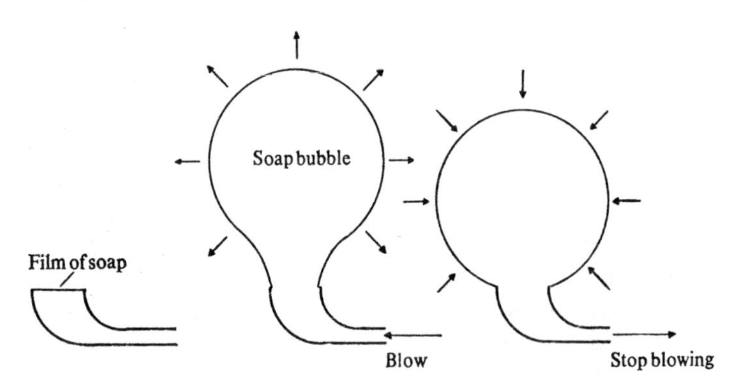

Fig.10.8. The blowing of a soap bubble and surface tension (see text).

it is once more a flat film of soapy water. No matter how many times you repeat the experiment, provided that the bubble does not burst, once you stop blowing the bubble will shrink. It is clear that unless you apply force the bubble has no stable existence so long as it has an outlet for the air it contains.

Why should this be? There is a force known as surface tension which exists because molecules of water attract one another. Putting it very crudely the water molecules like to be close-packed in crowds and dislike being pulled apart. What happens therefore when you blow a soap bubble? To start with, a thin film of soap water is stretched across a ring. When you blow, that film is stretched out more and more thinly and the water molecules are pulled further and further apart. They resist violently and only separate because they are forced apart by your blowing. When you stop blowing they rush together again and the surface area of the bubble shrinks until it is once more a flat film.

This is interesting, but why should it matter to a nurse? The lungs consist of millions and millions of tiny air sacs known as alveoli at the ends of a complex system of branching tubes whose function is to transfer air to and from the alveoli. It is in the alveoli that air comes into close contact with the blood: it is there that oxygen enters the blood and carbon dioxide leaves it. The alveoli are blown up and partially let down again about twenty times a minute in the process of breathing. Surface tension is important here because all the alveoli are lined and kept moist by a watery

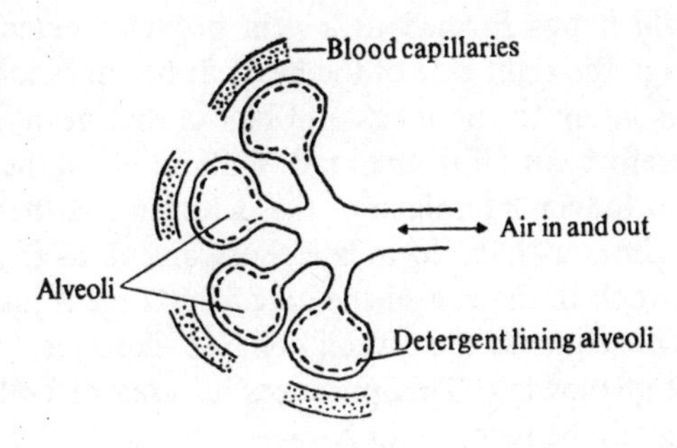

Fig.10.9. Because the alveoli are all lined with a fluid film, the effect of inflating the lungs is similar to that of blowing up millions of minute soap bubbles.

film. Breathing-in therefore is rather like inflating millions upon millions of minute soap bubbles. Part of the effort of breathing is expended in overcoming the surface tension of the fluid lining of the alveoli.

The work required is however not so great as it might be. This is because the cells of the alveolar walls secrete into the lining fluid a detergent-like substance which reduces the surface tension of the lining fluid to about one-tenth of what it would be if the fluid were pure water. The detergent reduces the attractiveness of the water molecules for one another and allows them to be pulled apart more easily. This enables the lungs to be expanded with much less effort than would be the case if the detergent were not present.

Lack of the detergent may be significant in several types of respiratory disease, but two are particularly important:

1. RESPIRATORY DISTRESS SYNDROME (HYALINE MEMBRANE DISEASE) OF NEW-BORN INFANTS. Some infants, especially premature ones, may start to breathe after birth but then very quickly get into trouble with their respiration. On watching them it is clearly apparent that they are having to struggle exceptionally hard to get air into their lungs; many of them give up and die. The precise cause of the syndrome is still uncertain, but when these infants are examined after death they are found to have virtually no detergent in their alveoli. As a result they must work about ten times harder than a

normal infant in order to breathe. Many of them cannot stand the strain and so die.

2. PULMONARY EMBOLISM. A pulmonary embolus occurs when a blood clot which has formed in a vein becomes detached and is washed through the right side of the heart. It becomes lodged in one of the arteries going to the lungs and blocks that artery. That area of lung is therefore cut off from blood. Without blood the cells of the alveoli can no longer manufacture the detergent. It therefore soon becomes ten times as hard to inflate those alveoli as it is to inflate the normal alveoli in the rest of the lung. The force is just not available and in consequence the alveoli collapse like soap bubbles into which you stop blowing. This produces an area of collapsed lung which can usually be seen on an X-ray.

11

Gases

It is important to have some idea of the basic properties of gases in order to understand both respiration and the oxygen cylinders and anaesthetic machines which every nurse sees in use every day.

The air

It will be as well to start by discussing the properties of the air with which we are surrounded. The air is a remarkably thin layer covering the earth: above a height of 20 miles or so there is virtually no air at all and even at a height of about 4 miles there is not sufficient air to sustain life. It is the column of air above us, pressing down upon the earth's surface, which produces the atmospheric pressure.

We do not notice this pressure because it is there all the time, but it is fairly easy to demonstrate its existence. Take a glass jar and immerse it in water until all the air has been expelled from it. Then invert the jar and slowly raise it from the water. The water will not fall out of the jar until the rim is actually lifted above the surface of the water. The water is held up in the jar because of the atmospheric pressure which is acting over the whole surface of the water apart from that protected by the jar. In fact if you had a long enough tube sealed at one end, you could demonstrate that the atmospheric pressure will hold up a column of water about 33 ft (about 10 m) high. Thirty-three feet of water is equivalent in weight to a column of mercury of equal diameter which is 760 mm high. (Mercury is 13·6 times as heavy as water.) Therefore at sea level the atmospheric pressure is said to be about 760 mmHg (Hg is the chemical symbol for mercury). Of course the precise pressure varies a little with the weather but it is never far from 760 mmHg. As one rises up a mountain or goes up in an aeroplane, the column of air above becomes shorter and shorter and so the atmospheric pressure becomes less and less.

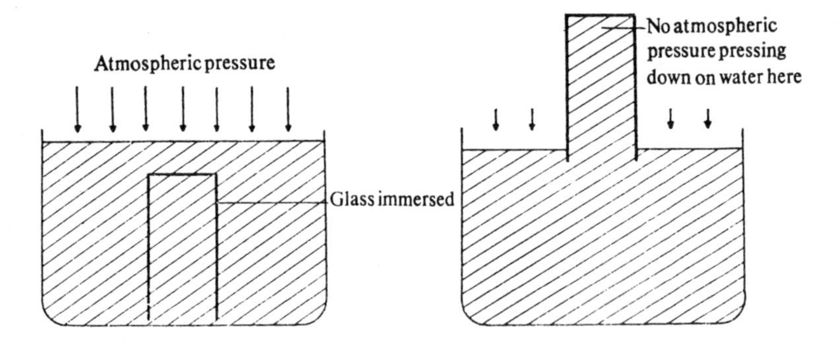

Fig.11.1. Demonstration of the existence of atmospheric pressure (see text).

The composition of the air

Air is made up of about 21 per cent of oxygen and about 79 per cent of nitrogen. Other gases such as carbon dioxide and the very rare gases like argon and xenon make up well under 1 per cent of the air.

This means that of the column of air stretching above you, 21 per cent consists of oxygen. Therefore, 21 per cent of the pressure which that column of air exerts is due to oxygen. Twenty-one per cent of 760 is 160 and so we say that the 'partial pressure' of oxygen in air at sea level is 160 mmHg. It is called partial pressure because it is that part of the total atmospheric pressure which is due to oxygen. The remaining 79 per cent of the air is nitrogen whose partial pressure at sea level is therefore about 600 mmHg.

The partial pressure is often written by using a small p and following it with the chemical symbol for the gas. Thus the partial pressure of oxygen can be written pO_2 and the partial pressure of nitrogen can be written pN_2. We use O_2 and N_2 rather than O and N because oxygen and nitrogen atoms each have a great affinity for their fellows. Pairs of atoms therefore tend to combine together giving gas molecules, each of which contains two atoms.

The partial pressure of a gas is roughly equivalent to its 'concentration'. We have seen that in solutions dissolved solids tend to move from areas where they are high in concentration to areas where they are low in concentration. Gases are similar in that they diffuse from areas where their partial pressure is high to areas where it is low.

As already mentioned, if one goes up to high altitudes the total atmospheric pressure falls. Suppose that one goes up in an aircraft

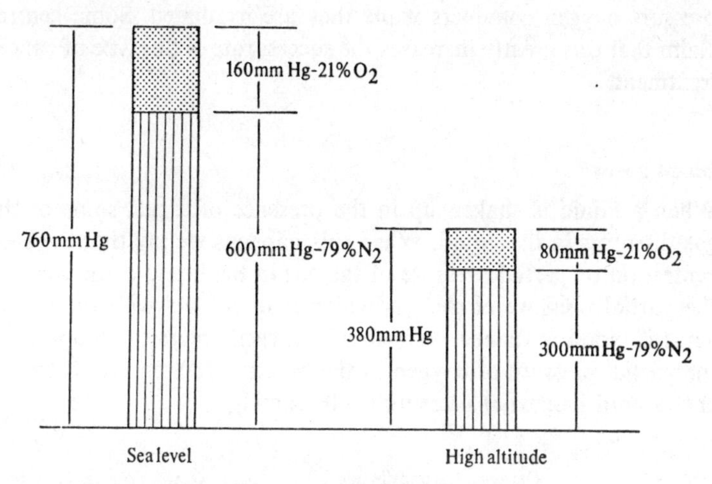

Fig.11.2. The composition of the air and the pressure of oxygen and nitrogen at sea level and at high altitude.

to a height where the total atmospheric pressure is just half that at sea level, namely 380 mmHg instead of 760 mmHg. At that altitude, 21 per cent of the air is still oxygen and 79 per cent is still nitrogen. But the partial pressure of oxygen is 21 per cent of 380 which is 80 mmHg. The 'concentration' of oxygen is therefore also half what it is at sea level.

In contrast, suppose that a patient is breathing pure oxygen from an oxygen cylinder. It is one of the laws governing the behaviour of gases that any gas released into the air will expand or contract until its pressure is the same as that of the surrounding air. Thus although oxygen is stored in the cylinder under high pressure, as soon as it is released from the cylinder it expands until its pressure becomes the same as that of the atmosphere, i.e. 760 mmHg. But the whole of that pressure is provided by oxygen and so the concentration of oxygen is about five times what it is in normal air.

Recently, experiments have been conducted on treatment with high pressure oxygen. The patient is put in a closed chamber which is pumped full of oxygen until its pressure is two or three times atmospheric pressure. The patient is thus exposed to ten to fifteen times the normal partial pressure of oxygen. One use for such chambers arises from the fact that cancers seem more susceptible to destruction by radiation when the tumour tissue is highly oxygenated. Sometimes patients who are being treated by radiotherapy are put in high pressure oxygen chambers while they are irradiated. Some centres claim that this greatly increases the success rate of this type of cancer treatment.

Blood gases

When a liquid is shaken up in the presence of a gas, some of the gas dissolves in the liquid. When this happens we say that the concentration or partial pressure of the gas in the liquid is the same as the partial pressure of the gas which is in contact with the liquid. Thus if blood is shaken up with air, oxygen enters the blood until the partial pressure of oxygen in the blood is 160 mmHg. Nitrogen enters until its partial pressure is 600 mmHg.

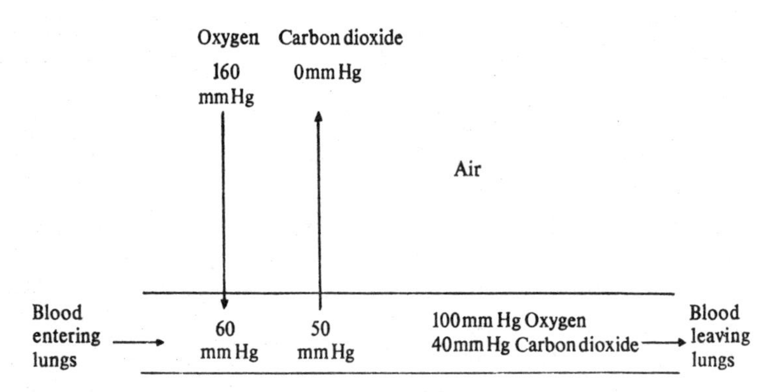

Fig.11.3. Changes in the composition of the blood as it flows through the lungs. Both oxygen and carbon dioxide diffuse because of differences in concentration (partial pressure).

The oxygen concentration of blood entering the lungs is variable but is often in the region of 60 mmHg. The partial pressure of carbon dioxide in this blood is usually in the region of 50 mmHg. The pO_2 of the outside air is about 160 mmHg while the pCO_2 in the outside air is virtually zero. There is therefore a gradient of concentration between air and blood which makes oxygen diffuse into the blood and carbon dioxide diffuse out of it. By the time the blood leaves the lungs enough oxygen has diffused into it to raise its partial pressure to about 100 mmHg, while enough carbon dioxide has diffused out to lower the pCO_2 to about 40 mmHg.

The body requires that the pO_2 in the blood leaving the lungs which will be pumped out by the left side of the heart should be about 100 mmHg. If there are abnormal barriers to the diffusion of gas as there may be if the alveoli are filled with fluid as in cardiac failure or pneumonia, the concentration gradient from 160 mmHg in outside air to 50–60 mmHg in the venous blood entering the lungs may not be sufficient to raise the pO_2 of arterial blood to 100 mmHg. The arterial blood instead of being fully oxygenated and bright red is only partially oxygenated and bluish and the patient is said to be cyanosed.

The situation can be greatly improved by increasing the concentration gradient. This can be done by making the patient breathe oxygen or an oxygen/air mixture instead of ordinary air. The air breathed in then has a much higher pO_2 than normal and the increased concentration gradient may be sufficient to enable the arterial pO_2 to be raised to normal levels.

Ordinary air is always relatively humid and so the rate of evaporation of water from the lungs and throat is kept low. But oxygen from an oxygen cylinder is quite dry and so if a patient breathes it fluid tends to be rapidly lost from the lungs and throat. The surfaces become dry and eventually may be damaged and infected. This means that if oxygen is to be breathed for any long period of time it is essential first to pass it through a humidifier. This usually consists of a device for bubbling the oxygen through water. In this way the gas becomes humid, the rate of evaporation from the throat and lungs is much less and the patient feels much more comfortable.

Pressure and volume

The relationship between the pressure and volume of a gas may be simply illustrated by the use of a syringe. If you take an ordinary 10 ml syringe and fill it full of air, so long as the end of the syringe is open the pressure within it will be the same as the pressure of the atmosphere outside. Now seal the end of the syringe with a finger and push the plunger in. You will have little difficulty in pushing it at least as far as the 5 ml mark. This demonstrates that gases are quite different from liquids in that they can be easily compressed.

Compression of the gas does however raise the pressure of the gas. You can feel this in a crude way by the pressure of the air on the finger which is sealing the syringe when the plunger is pushed in.

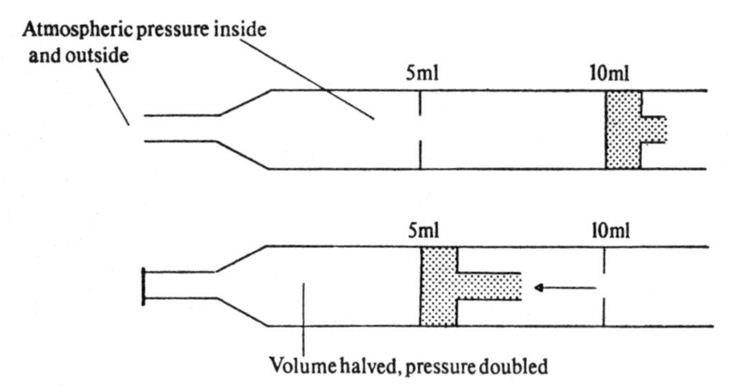

Fig.11.4. The syringe experiment showing the relation between pressure and volume in a gas (see text).

If you remove the finger which has been sealing the syringe, the higher pressure air inside shoots out until the air inside the syringe, like that outside, is once again at atmospheric pressure.

Now with the syringe plunger at the 5 ml mark, again seal the end with a finger and pull the plunger out to the 10 ml mark. The air inside is now having to occupy double the space and as a result its pressure is reduced to half. You can feel the effect of the reduction in pressure by the sucking sensation on the sealing finger. This time when that finger is removed, air rushes into the syringe until the pressure inside the syringe is again equal to the pressure outside.

These very simple experiments with an ordinary syringe illustrate the following important points about the behaviour of gases:

1. Gases can be compressed and when this happens their pressure rises. Conversely when a given amount of gas has to occupy a larger volume, its pressure tends to fall.

2. Gases flow from regions where the pressure is high to regions where the pressure is lower. The flow continues until the two pressures are equal.

The working of the lungs

The way in which air moves from the atmosphere into the lungs and back again can be easily understood from the simple laws governing the behaviour of gases which have just been outlined. First think of the lungs when you have just finished breathing out quietly. They still contain quite a lot of air and that air has the same pressure as the air in the atmosphere outside. Because of this the air is neither moving into nor out of the lungs.

Now what happens when you breathe in? The muscle known as the diaphragm flattens out and the chest wall expands, so greatly in-

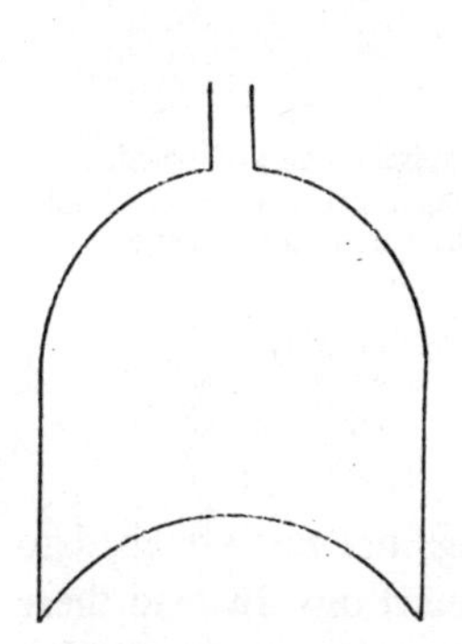

End of expiration, chest deflated, diaphragm high, pressure atmospheric inside and outside, no air movement

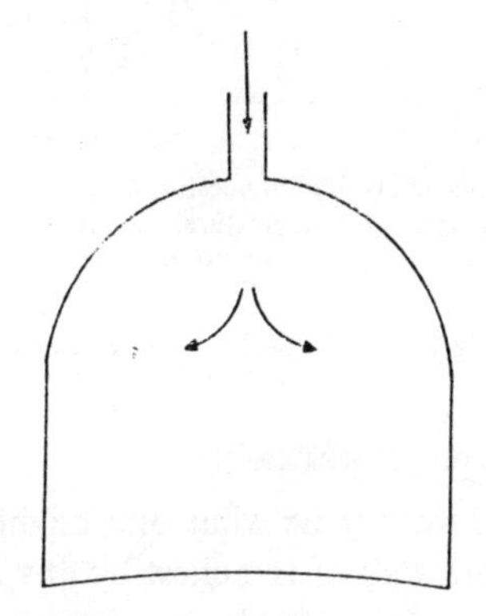

Inspiration, chest expanded, diapragm flat, pressure inside chest below atmospheric, air moves in

Fig.11.5. The inflation of the lungs (inspiration or breathing in).

creasing the volume of the lungs which must be filled by air. But initially the only air in the lungs is that which was left there after breathing out and which occupied a much smaller volume. When the chest expands, this air must fill a much larger volume and so its pressure falls. The pressure of air in the lungs therefore becomes lower than the pressure of air in the atmosphere outside and as a result air moves into the chest. The inward flow of air continues until the pressure inside the chest is again atmospheric.

Therefore at the end of breathing in, the expanded chest is full of air at atmospheric pressure. What happens on breathing out? The diaphragm rises and the chest wall relaxes, so reducing the volume of the lungs. The air in the lungs is thus compressed and so its pressure rises above the pressure of the outside air. Air then flows out of the chest until the pressure at the end of breathing out again becomes atmospheric which is the point at which this explanation started.

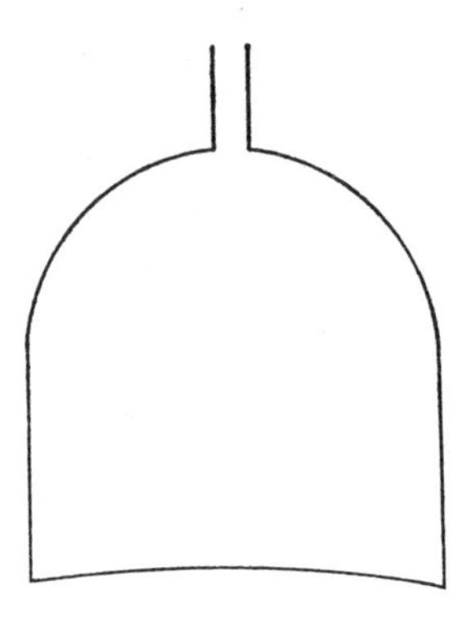

End of inspiration, chest expanded,
diaphragm flat, pressure inside chest
atmospheric, no air movement

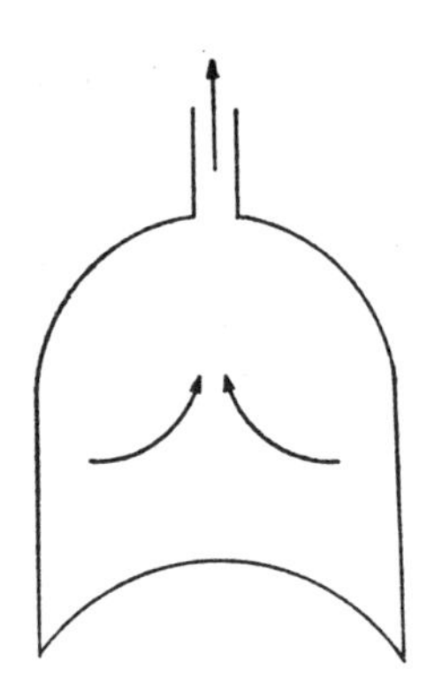

Expiration, chest deflated,
diaphragm high, pressure inside
chest above atmospheric,
air moves out

Fig.11.6. The deflation of the lungs (expiration or breathing out).

The pleural cavity

Contrary to what one might expect, the lungs are not attached to the wall of the chest by any actual tissue connections. Instead there is a remarkable arrangement which it is essential to understand if some of the diseases of the chest are to be properly appreciated.

The lungs themselves are covered by a very smooth, moist membrane known as the visceral pleura. The chest wall is lined by a similar smooth, moist membrane known as the parietal pleura. In normal individuals, these two layers of pleura are closely applied to one another. Although there are no actual bridges of tissue between them the two layers never normally come apart. When the chest wall expands, the visceral pleura is pulled outwards in company with the parietal pleura and the visceral pleura pulls out the underlying lungs. The pleural cavity is the space between the two pleural layers. Normally it contains nothing but a very thin film of liquid and it is thus more of a potential space than a real one in normal people.

What is the mysterious force which holds the two layers of pleura together so firmly without any anatomical connections? It is another example of the importance of surface tension which we discussed briefly in the last chapter. The only link between the two smooth, moist surfaces is a thin film of water. The water molecules in this film strongly resist being pulled apart from one another. Therefore when the parietal pleura moves outwards as the chest expands, it pulls on the water molecules which again pull on the visceral pleura. The two pleural layers are closely and effectively stuck together because of the existence of this thin, watery film.

The same force can be easily and convincingly demonstrated outside the body by using two microscope slides. Put a drop of water

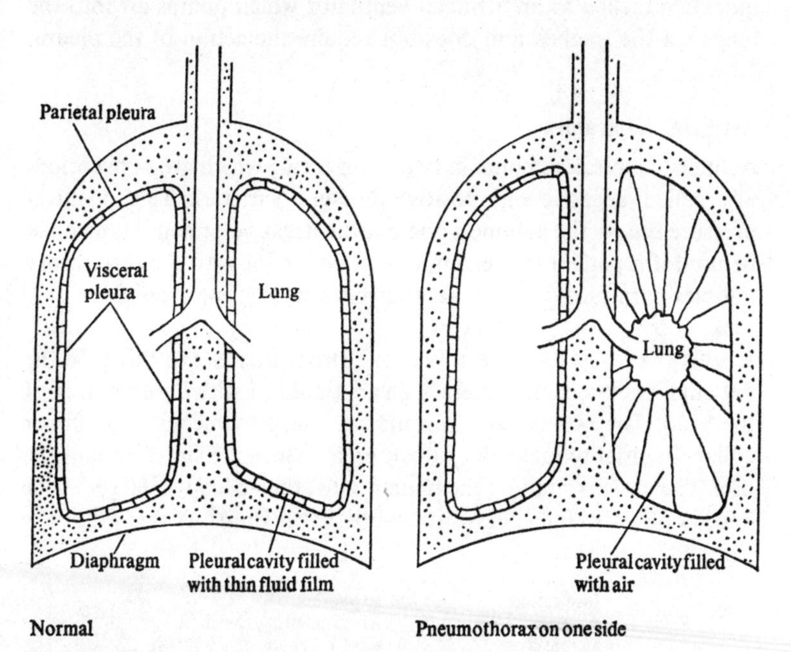

Fig.11.7. The lungs and pleural membranes showing how a pneumothorax can occur (see text).

on one slide and then press the second one down on top of the first. If you now lift up the top slide the bottom one will be lifted with it, held by the force of surface tension. Try to pull the slides apart and you will be surprised how powerful the force is.

The pleural cavity is important because sometimes, especially after a chest injury, one or other of the pleural layers is torn and air gets into the cavity either from the lungs or from the outside. This breaks the link between the visceral and parietal pleura. The lungs are in fact rather like balloons which are held inflated because they are attached to the chest wall by surface tension. When this link is broken the natural elasticity of the lungs causes them to collapse and most of the air is emptied from them. No longer then can the lungs expand and contract with the movements of the chest wall.

This condition is known as a pneumothorax and it is very distressing because even if it is confined only to one side, the other lung must then do all the work of breathing. If it occurs on both sides, both lungs collapse and death may occur quickly unless the patient is quickly attached to an artificial ventilator which pumps air into the lungs via the trachea and does not require the action of the pleura.

Artificial respirators

Artificial respirators come in two main forms which are conventionally called negative-and positive-pressure varieties. The positive-pressure type is the common one but the negative-pressure type must be used if a patient (for example a polio victim with the respiratory muscles paralysed) is to be kept alive for months or even years.

1. NEGATIVE-PRESSURE TYPE OR 'IRON LUNG'. In this type the patient is enclosed completely in an air-tight box with the exception of the head. The box is sealed around the neck by means of a rubber collar. In order to make the patient breathe in, air is sucked out of the box. The air surrounding the patient must then occupy a larger space and its pressure falls below atmospheric pressure. At this point the

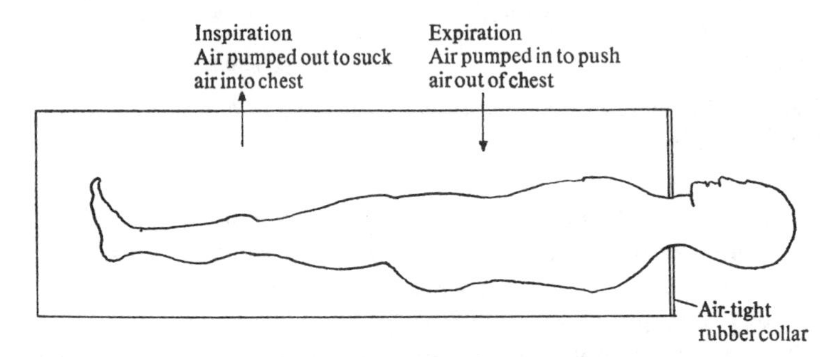

Fig.11.8. A negative-pressure artificial respirator (iron lung).

patient's lungs are filled with air at atmospheric pressure while his chest is surrounded by air at below atmospheric pressure. This pressure difference pushes the chest wall outwards and the lungs move with the chest wall. The pressure of the air inside the lungs therefore falls and air moves into the chest from the atmosphere until the pressures inside the chest and in the atmosphere are the same.

In order to achieve breathing out, air is pumped into the box around the patient. This pushes the chest in, the pressure of the air inside the chest rises, and air moves out of the chest again until pressures in the chest and in the atmosphere are equal. The iron-lung type of respirator thus produces a pattern similar to normal breathing, in that during inspiration the pressure in the chest is below atmospheric while during expiration it is above atmospheric: the air is sucked in and pumped out.

2. POSITIVE-PRESSURE TYPE. This is the common type of respirator used in operating theatres for anaesthesia and in the wards for relatively short periods of artificial respiration. A tube is placed in the patient's trachea, either directly via a tracheostomy at the front of the neck or by passing the tube down through the mouth or nose. Once the tube is in the trachea a little balloon around the lower end of the tube is inflated, so making an air-tight seal: all the air passing into and out of the lungs must then pass through the tube and none

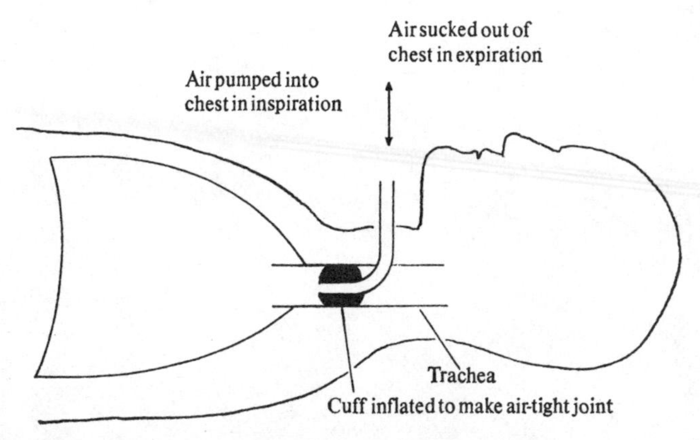

Fig.11.9. The action of a positive-pressure artificial respirator.

can pass around its sides. The tube is then connected to a pump which alternatively gives a pumping and sucking action. In order to make the patient breathe in, air is pumped into the chest at above atmospheric pressure: this forces air into the chest and the chest walls expand. Thus, during breathing in, the pressure in the lungs is above atmospheric, the opposite of the situation in a normal person and in someone being artificially respired by means of an iron lung. In order to make the patient breathe out, very gentle suction is applied. This lowers the pressure inside the chest and the chest walls contract, so pushing out the air: again this is the opposite of the normal situation.

Negative-pressure respirators require normal and intact pleural membranes if they are to work: they cannot therefore be used satisfactorily in patients with chest injuries. On the other hand positive-pressure respirators do not depend on the pleural membranes for their action and so they can be used to inflate the chest in patients who have pleural tears and a pneumothorax. The main disadvantages of positive-pressure respirators are that the tubes are often distressing and uncomfortable for the patient and unless very carefully handled the pressure of the balloon may damage the tracheal wall by interrupting the circulation. It is essential to let the balloon down for short periods every 2 hours or so in order to reduce the risk of damage.

12

Electricity

Electricity, like heat, is best understood in terms of the effects it has. This is especially true if one is interested not so much in electricity itself but in the applications it has to the understanding of medical science. Yet again this is not a chapter for physicists. But I hope it will be successful in its aim, which is to be helpful to nurses.

The behaviour and effects of electricity

Electricity, or electrical charge as it is often called has the following main characteristics:

1. It can only travel in certain materials known as conductors. The most important conductors are metals and water. Conductors vary considerably in the ease with which electricity can travel through them. Copper is one of the best conductors among the metals and that is why wires are often made of copper. Water conducts most effectively when it has dissolved in it substances which can give rise to ions. A salt solution, full of sodium and chloride ions, is a much better conductor than tap water.

2. Some materials are so resistant to the passage of electricity that they are known as insulators. Wood, plastic, dry cloth, rubber and air are all good insulators.

3. Electrical charge can be manufactured by a number of devices such as batteries, accumulators, dynamos and generators. It can be stored in devices known as capacitors.

4. The concentration of electrical charge at any point is known as

the electrical potential at that point. It is roughly equivalent to the pressure of a gas or of a liquid.

5. When two points carrying electric charges are linked by a conductor, charge always flows from the point where it is high in concentration to the point where its concentration is low. In other

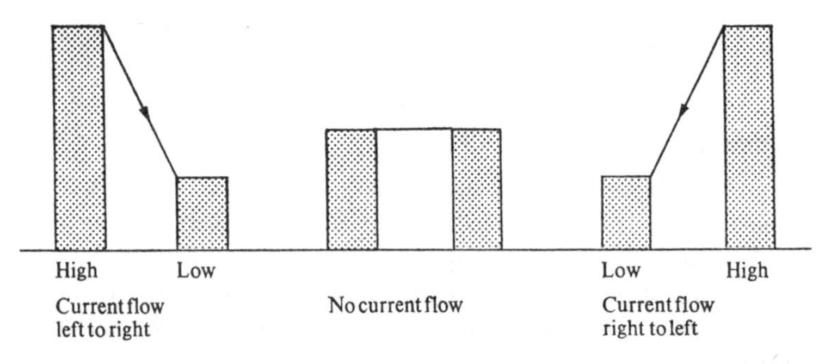

Fig.12.1. When two points are linked by an electrical conductor, current always flows from the point at high potential to the point at low potential. When two points are at the same potential there can be no current flow.

words, whenever there is a potential difference between two points and those points are linked by a conductor, charge always flows from the place where the potential is high to the place where it is low. A point of high potential is conventionally said to be positive to a place of low potential. Conversely a point of low potential is conventionally said to be negative to a point of high potential. Electrical charge is therefore usually said to flow from positive to negative points.

6. The rate of flow of charge in a conductor is called the current in that conductor. It is analogous to the flow of gas or liquid in a tube.

7. When two points are linked by a good conductor, charge can easily flow from one to the other and therefore any potential difference between two such points must be very small. This is similar to the situation in liquids where if a wide pipe links two points there can be little pressure difference between them. But if two points are linked by a poor conductor which resists the passage of electricity, then charge builds up in front of the resistance and the potential difference between the two points will tend to be high. Again this is similar to

the pressure in a water pipe before and after a narrow point which offers considerable resistance to flow.

8. The ease of conduction depends partly on the material of which a conductor is made and partly on the physical size of the conductor. As a wire increases in diameter, the resistance it offers to the flow of electricity falls. Again this is similar to the flow of water in a tube.

9. There are two opposite sorts of electricity known as positive and negative. They are similar in every way, apart from the fact that they can cancel one another out. The smallest unit of positive electricity is the charge on a proton. The smallest unit of negative electricity is the charge on an electron. Positive charges repel other positive charges but attract negative charges. Negative charges repel other negative charges but attract positive charges.

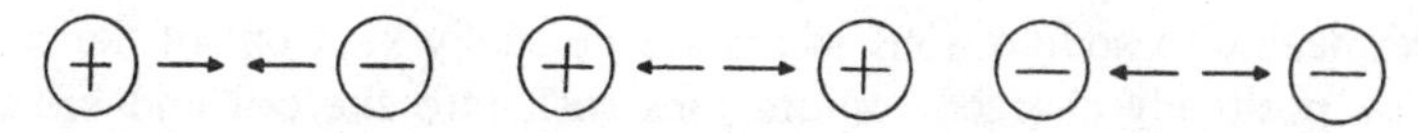

Fig.12.2. Like-charges repel one another, unlike-charges attract one another.

10. When electrical charge flows through a conductor, heat is given off. The amount of heat produced depends on two things, the amount of charge flowing and the resistance the conductor offers to that flow. The greater the flow of charge and the greater the resistance, the more heat is given off. Both electric lights and electric heaters depend on this property of electricity.

11. When electricity flows through a conductor it converts that conductor into a magnet. By arranging the conductors in special ways, powerful magnets can be created. These have many uses, the most important of which are in electric motors in which the rotating parts are pulled round by electrically generated magnetism.

12. Units of electricity. The potential difference between two points is measured in volts, the rate at which current flows through a conductor is measured in amps and the resistance of a conductor to current flow is measured in ohms. The power of a generator of electricity (the amount of charge it can generate in a given time) and the power consumption of a light, a heater or an electric motor (the amount of charge it uses in a given time) are both measured in watts.

Electricity in the body

Electricity appears to be extremely important in the normal functioning of the body. The muscles, the nervous system and the heart all depend upon it and can generate it. In fact many of the early experiments on electricity performed about two centuries ago used the electricity generated by the muscles of frogs.

The membranes of all cells appear to be devices which can pump electrical charges in the form of charged ions from one side to the other. As a result of this pumping activity, in most cells the inside is electrically negative to the outside. This is especially true of nerve and muscle fibres where the maintenance of this potential difference seems to be essential in order to keep the cell ready for instant activity.

The main feature of nerve activity is a brief change in the membrane which lasts not much more than a millisecond (one thousandth of a second). During this time the membrane suddenly becomes highly permeable to sodium ions which are normally kept out of the cell. The positively charged sodium ions rush into the cell and briefly make the inside of the cell positive to the outside.

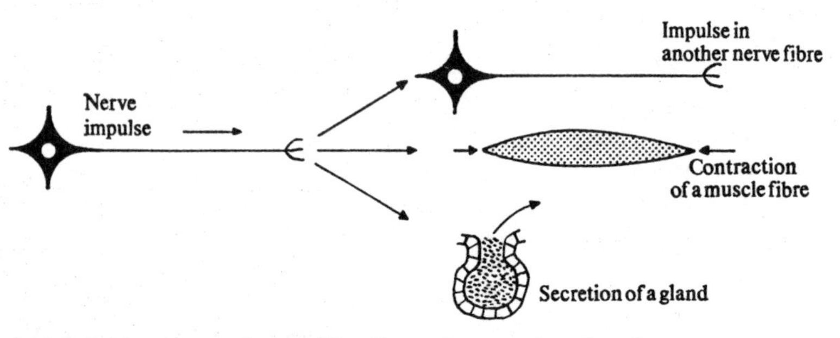

Fig.12.3. The three main possible effects of a nerve impulse.

Such an electrical disturbance is known as a nerve impulse and it can rapidly spread along a nerve fibre or over the surface of a muscle cell. When an impulse reaches the end of a nerve fibre it releases a chemical which may do one of three things:

1. Fire off an impulse in another nerve cell.
2. Fire off an impulse in a muscle fibre.
3. Stimulate a gland to secrete.

Impulses which travel over the surfaces of muscle fibres stimulate the contraction mechanism inside the fibre which then shortens the fibre.

The electrical disturbances produced by the normal heart and brain, although very small, can be picked up by sensitive recording instruments. An instrument known as an electroencephalograph (EEG; electrocorticograph or ECG in America) can record the electrical activity of the brain by means of devices known as electrodes placed on the skull which pick up the tiny electrical disturbances and feed them into a machine which amplifies (magnifies) them greatly and records them on paper. The normal pattern of electrical discharge is known from an enormous amount of past experience in the recording of EEGs from normal people. Abnormal patterns of electrical activity may appear in parts of the brain in patients with epilepsy and in those with brain tumours. The EEG may therefore be helpful in making a diagnosis.

The instrument which records the electrical activity of the heart is the electrocardiograph (ECG; or in America, EKG). The electrical discharges produced by the heart muscle are much larger than those produced by the brain. They are conducted throughout the body fluids and they can be picked up by electrodes attached to the wrists and ankles as well as by ones on the chest wall itself. Again, long experience has enabled the normal patterns and the abnormal ones caused by such conditions as coronary thombosis and prolonged high blood pressure to be worked out. The ECG is therefore an extremely useful aid in the diagnosis of heart disease.

Electric shock

Electrocution is one of the commonest forms of death and it is a particularly important killer of young people and children. A person is killed by electricity because the currents flowing through the body give a shock to the heart, stopping it beating, and a shock to the brain which stops the function of the part which controls breathing. Very severe shocks may produce permanent damage to these organs and full recovery is then impossible. Quite often, however, the damage is only transient and both heart and brain may quickly recover spontaneously. Also quite often, the damage, although temporarily severe, is not fatal provided that artificial respiration can be started at once to keep the breathing going. The heart may sometimes be started by thumping hard on the chest wall or oddly enough by giving a second shock to the chest by means of a machine known as a defibrillator.

The thing about electricity which kills is the amount of current which flows through the body. Two factors determine the magnitude of this current flow.

1. The potential (voltage) of the source of electricity to which the body is connected. The higher the potential, the greater will be the potential difference between the source of electricity and the body, and the greater will be the current flow. A 12 V battery will never send enough current through the body to kill. A high tension cable carrying thousands of volts will almost certainly kill. The ordinary 100–250 V of the mains supply (the precise level depends on the country in which you live) will quite often drive enough current through the body to kill but also quite often the patient will survive. With shocks from the mains the crucial question which determines the outcome is usually the next factor.

2. The resistance which the body offers to the flow of electricity. If this is very high, even sources of very high potential may fail to send enough current through the body to kill. If it is very low, quite weak electrical sources may produce a killing current. It is important to realise that in order to do its damage the current must flow through the body and out the other side. The resistance of the outflow from the body is therefore as important as the resistance of the inflow. This means that people wearing rubber shoes and gloves, standing on a dry rubber, wood or plastic floor are most unlikely to get a dangerous shock: the resistances of inflow and outflow are so high that even touching a high tension cable might not send enough current through the body. Dry skin also has a high resistance and so touching a live mains source with a dry hand is not very likely to kill. But skin that is moistened by sweat or water is quite a good conductor and if a wet hand touches the mains supply a dangerous shock is very likely. This is especially so if the feet and floor are wet also, hence the extreme danger of electrical applicances in bathrooms.

The mains and earthing

Most electrical machines used by nurses require power which can be tapped from the electrical mains supply by the use of a plug. Most such plugs now have three pins, two of which are essential for running the machine. The third, the earth or ground, is a safety device. The two pins required for running the machine are known as positive and negative or more usually as live and neutral (L and N).

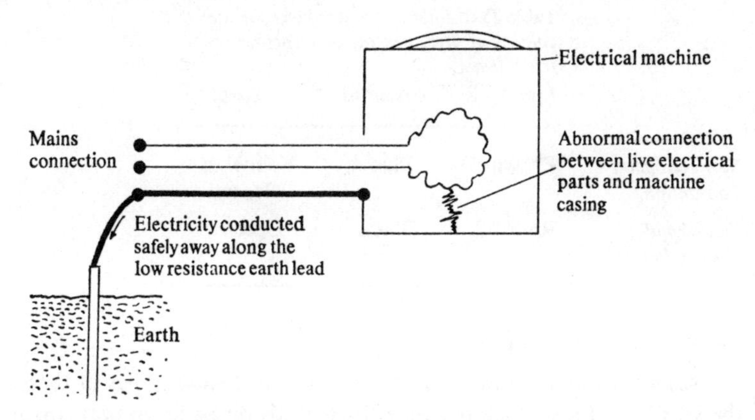

Fig.12.4. The action of the earth lead (see text).

The positive or live terminal is the source of the high potential: this drives the current through the machine which it then leaves by flowing out through the negative or neutral terminal.

The possession of an earth or ground terminal is an important safety device. At the machine end, the earth is connected to the casing and to any parts of the machine which could be touched by anyone using the machine normally. The earth terminal in the socket is eventually connected by a low resistance wire to a metal rod embedded deep in the ground. The point of the earth lead is therefore to provide a low resistance connection between the casing of the machine and the soil. If because of some fault in the machine the casing became connected to the live electricity this could clearly be highly dangerous to anyone handling it. The earth lead removes this danger because the electrical charge is conducted away down the low resistance wire. The earth lead is always of very much lower resistance than the human body and since electricity, like water, tends to flow preferentially along a channel where the resistance is least, the earth removes the danger of a severe shock.

There are a number of conventions in use for the colours carried by the coverings of live, neutral and earth wires and two of these are shown in the table. (N.B. This table shows the new European conventions and the traditional conventions only: other countries may use different conventions and it is important to learn the correct ones for your own country.)

Table *Two different conventions for the colours of live, neutral and earth electrical leads*

	Live	*Neutral*	*Earth*
New European Convention	Brown	Blue	Yellow & Green
Traditional Convention	Red	Black	Green

Electrical recording machines

Nurses are likely to come into contact with two main types of these, the ECG machine which produces its records on paper in permanent form and the cardiac monitor which displays the ECG on an oscilloscope screen. In the ordinary electrocardiograph, the positive and negative electrical activity picked up from the body is used to deflect a light pen. Positive electricity deflects the pen one way and negative electricity deflects it the other. The pen writes on paper moving underneath its tip, thus producing a permanent recording of the electrical activity of the heart. Some recorders actually use ink but in most of the newer types heat-sensitive paper is employed. This gives rise to a black mark whenever it is touched by a hot object. The tip of the pen is heated and as it moves over the paper it produces a recording which is much less messy than the ink type.

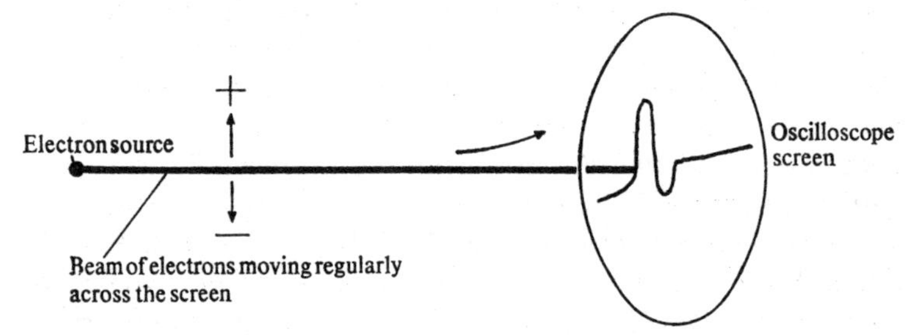

Fig.12.5. The principles on which ECG machines and oscilloscopes work (see text).

An oscilloscope is a much more sensitive method of detecting electrical activity. Essentially it consists of a mechanism for generating a fine beam of electrons which when they hit a special screen produce a glowing image. Depending on the material of which the screen is made, the image may disappear virtually instantaneously or it may persist for many seconds.

The electron generator is so arranged that the beam regularly sweeps across the screen from the observer's left to his right. It is usually possible to alter the speed of the sweep so that it takes anything from a fraction of a second to many seconds. The beam of electrons acts as an extremely light 'pointer'. Since it consists of a stream of negatively charged particles, it is attracted by positive electricity and repelled by negative electricity. Upward deflections often indicate a positive electrical signal and downward deflections a negative one. Thus an ECG can be displayed on an oscilloscope screen where it can be readily watched by a nurse or doctor.

Most modern machines for cardiac monitoring also have a counter which is triggered off by the electrical changes occurring during each heart beat. The counter can be so arranged that it triggers a bell or a flashing light if the heart rate should rise too high or fall too low.

Static electricity

Many types of cloth and plastic when rubbed, rustled or moved can generate electrical charge which builds up on the surface of the material. Because it does not flow anywhere but stays stationary, this is known as static electricity. Quite large concentrations of charge can be built up in this way and so the potential may become high.

High potentials of this type may, when a point of low potential comes close, cause a spark to jump from the point at high potential to the point at low potential. In this way the charge of static electricity is dispersed. Tiny sparks of this type are familiar to everyone as the crackles one hears when taking off a nylon shirt, sweater or underclothes. Because the generating source is usually weak such sparks are not usually dangerous unless the potential build up is enormous, as it can be in a thundercloud. However, even tiny sparks can be a menace in the vicinity of explosive gas mixtures as may be used in some types of anaesthesia. This is why many items for use in operating theatres are manufactured with the aim of reducing the likelihood of the accumulation of static electricity. Such items, and in particular boots, may often have the words 'Anti-static' stamped upon them.

13

Units, decimals and graphs

Most hospitals in most parts of the world now use the metric system originally devised in France nearly 200 years ago. In this system the prefix 'milli' means one thousandth and the prefix 'micro' means one millionth. The most important units in the metric system are as follows:

Length 10 millimetres (mm) = 1 centimetre (cm)
100 centimetres (cm) = 1 metre (m)

One metre is approximately 39 inches or just over 1 yard.

Volume 1,000 millilitres (ml) = 1 litre (l)

One millilitre is also equivalent to 1 cubic centimetre (cm^3). One litre is roughly 1·8 pints.

Weight 1,000 micrograms (μg) = 1 milligram (mg)
1,000 milligrams (mg) = 1 gram (g)
1,000 grams (g) = 1 kilogram (kg)

One gram is the weight of 1 cubic centimetre or 1 millilitre of water. One kilogram is therefore the weight of 1 litre of water. One kilogram is equal to 2·2 pounds.

Decimals

It is increasingly common for drug dosages to be expressed in decimals rather than fractions. In some ways this is a pity because it is far easier to make serious mistakes when decimals are used. The essential thing to remember is that the position of the decimal point must be noted very carefully. If you do not do this it is all too easy to give a dose which is ten times too great or ten times too little. If the writing is unclear and you are not absolutely certain where the decimal point comes it is essential to ask the person who wrote the figures down. Already there have been serious accidents because of a combination of bad writing by doctors and an unwillingness on the part of nurses to check with the doctor that the point is in the right position. Some examples of decimal notation follow:

One	1	1·0
One-tenth	1/10	0·1
One-hundredth	1/100	0·01
One-thousandth	1/1000	0·001
One-quarter	1/4	0·25
One-eighth	1/8	0·125
One-third	1/3	0·333

Graphs

A graph is a way of recording information so that it can be instantly taken in after only a quick glance. Most graphs in medicine are used to demonstrate the way in which some factor changes with time, and the two factors encountered most frequently by nurses are temperature and blood pressure.

In drawing such a graph time is always plotted on the horizontal scale and the factor which is being measured is always plotted on the vertical scale. This produces a clear picture of how something is changing with time and this can be appreciated much more easily than when all the figures are put in a table or written down in the patient's case notes.

Index

NOTES

NOTES

NOTES

NOTES

NOTES